WITHDRA
UTSA LIBRARIES
D0436244

THE CHICAGO GUIDE TO
COMMUNICATING SCIENCE

On Writing, Editing, and Publishing
Jacques Barzun

Getting into Print
Walter W. Powell

Writing for Social Scientists
Howard S. Becker

Chicago Guide to Preparing Electronic Manuscripts
Prepared by the Staff of the University of Chicago Press

Tales of the Field
John Van Maanen

A Handbook of Biological Illustration
Frances W. Zweifel

The Craft of Translation
John Biguenet and Rainer Schulte, editors

Style
Joseph M. Williams

Mapping It Out
Mark Monmonier

Indexing Books
Nancy C. Mulvany

Writing Ethnographic Fieldnotes
Robert M. Emerson, Rachel I. Fretz, and Linda L. Shaw

Glossary of Typesetting Terms
Richard Eckersley, Richard Angstadt, Charles M. Ellerston, Richard Hendel, Naomi B. Pascal, and Anita Walker Scott

The Craft of Research
Wayne C. Booth, Gregory S. Colomb, and Joseph M. Williams

A Manual for Writers of Term Papers, Theses, and Dissertations
Kate L. Turabian

Tricks of the Trade
Howard S. Becker

A Poet's Guide to Poetry
Mary Kinzie

Legal Writing in Plain English
Bryan A. Garner

Getting It Published
William Germano

THE CHICAGO GUIDE TO COMMUNICATING SCIENCE

SCOTT L. MONTGOMERY

THE UNIVERSITY OF CHICAGO PRESS
CHICAGO AND LONDON

Scott L. Montgomery is a consulting geologist, writer, and independent scholar. He is the author of several books on the history of science and scientific language, most recently *The Moon and the Western Imagination* and *Science in Translation: Movements of Knowledge through Cultures and Time*.

The University of Chicago Press, Chicago 60637
The University of Chicago Press, Ltd., London

Printed in the United States of America

12 11 10 09 08 07 06 05 04 03   1 2 3 4 5

ISBN: 0-226-53484-7 (cloth)
ISBN: 0-226-53485-5 (paper)

Library of Congress Cataloging-in-Publication Data

Montgomery, Scott L.
The Chicago guide to communicating science / Scott L. Montgomery.
p. cm. — (Chicago guides to writing, editing, and publishing)
Includes bibliographical references and index.
ISBN 0-226-53484-7 (alk. paper) — ISBN 0-226-53485-5 (pbk. : alk. paper)
1. Communication of technical information. 2. Communication in science. 3. Technical writing. I. Title. II. Series.
T10.5 .M65 2003
501′.4—dc21
2002072627

♾ The paper used in this publication meets the minimum requirements of the American National Standard for Information Sciences—Permanence of Paper for Printed Library Materials, ANSI Z39.48-1992.

To Kay and Shirley

CONTENTS

PREFACE

This book is a product of much time and labor spent in the halls, towers, moats, and dungeons of scientific communication. My own work as a publishing scientist and as an independent scholar in the history of science and scientific language has provided much material to draw upon. But this work is only part of the story. Like a great majority of writers, I've been both peasant and yeoman in the fields of contemporary authorship. This service has entailed being urged or dragged into many roles: essayist, freelancer, translator, scriptwriter, speaker, critic, reviewer, ghostwriter, copywriter, editor, fix-it boy, messenger, secretary, and (not least) rejectee. These are the many faces of the "successful" writer in contemporary society—the writer, that is, whose work is allowed to see print on something approaching a regular basis. The chapters that follow come directly from such multipronged experience.

This book is quite likely to differ from most other guides to scientific communication you might come across. The reason is that, at base, I treat scientists not as literary underdogs—that is, as scientists *first,* who must also sometimes (somehow, against whatever odds) express themselves intelligibly—but instead as full-fledged writers or speakers, who understand that the transfer of their knowledge to others is part of the essence of research. Writers, in particular, learn to write most of all not from paternal rules and standards, nor from the mother's milk of step-by-step advice. They learn from their brethren, from other writers—the Latin term *imitatio,* which evokes the combined sense of imitation, adaptation, and invention, is a good one to apply and, in fact, derives from the rhetorical tradition, from ancient to modern times, of teaching composition. Learning from other writers, however, is not as simple as it may sound. There are a number of important and practical aspects to it that I have sought to integrate, at various levels, throughout this book.

The result is a guide that views the scientist, in part, as a species of the genus "writer" (or rather, "communicator"). It has been my intent to offer a series of clear and realistic choices for how to develop skill at either a functional or a superior level, on a personal basis, and in direct awareness of the realities of writing, speaking, and publishing in science today. At the same time, part of this book is also aimed at teaching the scientist-author something about the nature and history of his or her discourse, as a living, evolving phenomenon. Knowledge of the medium, as well as how to wield it effectively, can only give one a degree of added control over it. And indeed, such control is really the crux of the matter.

Noam Chomsky, quite likely the greatest linguist of the past half century, once called the dictionary a "list of hints." His point was that words and meanings are too pliable in their uses to be held prisoner by a fixed set of definitions. Similarly, those who hope to find the eternal, unchanging rules of good scientific expression are bound to drink from many fountains while still growing old. The flow of science, in form and articulation, has always been diverse, offering reasons for writers and speakers to both adapt and explore.

The writing of this book has benefited greatly from help supplied, over time and not always intentionally, by many friends, teachers, and colleagues. I have space here to mention only a few to whom I owe the deepest gratitude. Bill Travers, geologist and friend, formerly of Cornell University, encouraged many useful discussions over the years on science and writing. Nigel Anstey, geophysicist *magnum,* helped reorient my thinking about certain important subjects. During the past five years, I've received the benefit of helpful criticism from the many reviewers and editors at the American Association of Petroleum Geologists, the Geological Society of America, the Kansas Geological Society, the Rocky Mountain Association of Geologists, the Utah Geological Survey, and the U.S. Geological Survey, to name only the most prominent sources of obligation. Useful exchanges and forums for talking about scientific language and publication were provided, along the way, by Steve Fuller, Les Levidow, and Kirk Junker, with often quiet but nonetheless effective support from John Lyne. Tom Cottner, of the University of Washington Medical Research Center, provided valuable material and information for the present volume. I would also like to thank the reviewers of the original manuscript, who, by their intelligence and knowledge, improved many sections of the book and saved me from certain mistakes of information and judgment. Any remaining errors are entirely my own.

Responsibility for making this book a reality and for ensuring that it retains whatever quality it might possess must be given also to Susan Abrams of the University of Chicago Press. Her unflagging support, patience, and intelligent suggestions during all phases of the project will always be remembered.

Finally, I offer this work as homage to family members, living and passed on, especially Kay, Shirley, Lynn, and Frank, all of whom would have expressed surprise to see such a book emerge from these hands. Last, I must thank those "within the palisade"—Kyle, Cameron, and Marilyn—who have endured yet another season of hopeful labor.

CHAPTER ONE

Communicating Science

Whoever tells the truth, sooner or later will be caught doing it.
—Oscar Wilde

FIRST THINGS

Science exists because scientists are writers and speakers. We know this, if only intuitively, from the very moment we embark upon a career in biology, physics, or geology. As a shared form of knowledge, scientific understanding is inseparable from the written and spoken word. There are no boundaries, no walls, between the doing of science and the communication of it; communicating *is* the doing of science. If data falls in the forest, and no one hears it . . . Research that never sees the dark of print remains either hidden or virtual or nonexistent. Publication and public speaking are how scientific work gains a presence, a shared reality in the world.

These basic truths form a starting point. As scientists, we are scholars too, steeped in learning, study, and, yes, competitive fellowship. Communicating is our life's work—it is what determines our presence and place in the universe of professional endeavor. And so we must accept the duties, as well as the demands and urges (and, fortunately or unfortunately, the responsibilities), of authorship. But aside from noble sentiment, there are other reasons for trying to communicate well with our intellectual brethren.

No one who aspires to a scientific career can afford to overlook the practical implications of what has just been said. The ability to write and speak effectively will determine, in no uncertain terms, the perceived importance and validity of your work. To a large degree, your reputation will rest on your ability to communicate. The reason to improve your skill in this area, therefore, is not to please English teachers past and present (though these may well haunt us till we shed our mortal coil). It is to gain something very real in the professional world, something of advantage. To communicate well is to engage in self-interest. Another way of saying this is that writing and speaking intelligibly are required forms of professional competence—nothing less.

Contrary to what you may feel, however, this situation is not a fatal one. Creating and sharing knowledge are truly profound but also eminently performable acts. While they are among the highest achievements of which human beings will ever be capable, they are done every day. If they weren't, science would cease to exist. Every time you put pen to paper (or finger to keyboard), step up to the podium, or clear your throat in front of a class, you become a full participant in what has clearly become humankind's most powerful domain of intellectual enterprise.

The purpose of this guide is to help you, the scientist, deal competently, even eloquently, with your role as an author. My intent is to aid you in learning how to feel at home with, and even take significant pride in, the communicating you will do as a member of the greater scientific community. This can be done, as it happens, without torture or torment, golden rules or iron systems. What it does require, among other things, is patience, a willingness to learn from others, and a certain way of looking at authorship.

THE IMPORTANCE OF ATTITUDE

Writing, we know, does not always come easily to scientists. Innumerable tales can be told of brilliant researchers whose papers would blind the eye of a first-year composition instructor. Yet, in reality, good writing rarely comes easily to *anyone,* in *any* discipline, whether quantum mechanics or art history. Writing is aptly called a skill, or, more accurately, a collection of skills. It is never entirely mechanical and always involves a level of emotional engagement, as well as forbearance and discipline. The Japanese have an excellent proverb for what it takes to learn a skill: "Ishi no ue ni, jyū nen." Ten years, standing on a rock.

I'm not suggesting that we try this (three to five years, with time off for good behavior, should be plenty). But it points in a certain direction. What has our training, as scientists, been like in this area? In fact, a major difference between the humanities and sciences is that composing, critiquing, and revising papers forms a central part of learning in the former, while in the sciences it does not. Moreover, immersing oneself in eloquent writing of the past is also prominent in humanities training, whereas scientific instruction tends to avoid this sort of thing almost entirely. We don't read Newton (or much of him) in a basic physics class, Linnaeus in a botany course, Lavoisier or Lyell in a chemistry or geology curriculum. Why is this so? The reasons are complex, and have much to do with the recent history of science. But the effects are clear: good writing is something that scientists are supposed to pick up, either from a course or two in technical writing while in school, or through osmosis after entering the caffeine-ridden world of professional research.

If formal communication can be intimidating for scientists and engineers, what is the best way to help gain the upper hand? Much begins with how one thinks about writing in particular and about scientific language in general. To communicate well, you need to feel at least some degree of *control* over the language you are using. This means a basic awareness that you, the writer, are taking words and images and creating something out of them. It also means an understanding that you are doing this by employing certain forms and structures toward the goal of persuading—telling a story to—a very particular kind of audience.

Too often in science we have the feeling that our language is not really our own. Technical speech seems something largely prefabricated, consisting of hardened and formal pieces that need to be chosen, organized, and intelligently assembled. There is a drop of truth here; scientific writing *is* generally flat, heavily reliant upon preexisting technical terms and phrases. Journal editors are unlikely to smile favorably at literary turns of phrase, passionate outbursts, or fanfares to the gods of invention. Yet this hardly describes the whole of the matter. Science may sound impersonal to the ear, but it is fully human and personal to the touch. The calm, declarative "voice" of technical speech is something we must make anew, every time, through a host of choices, a number of which are actually quite flexible. If we look closely enough, we can find many avenues where personal eloquence may be put to practical use. The creative and the individual have a very important dimension in our writing (I'll say more about this in chapter 4).

Much begins and ends with attitude. Reasonably confident authors transfer their sense of self to the reader. Their science tends to be effective, less hesitant. If you are terrified of writing, it is likely that your writing will terrify others (or worse, inspire humor). Conversely, if you view the composition of technical papers as an unbounded creative exercise, with enthrallment as its goal, you will meet a quick and scarlet end at the hands of the first editor to come along. This book has been written to protect you from both fates.

THE EXISTING LITERATURE ON TECHNICAL COMMUNICATION: A MILITARY REVIEW

I would be remiss, both as a scientist and as a writer, if I did not include some pointed words about my competitors. In technical terms, this means a "review of the existing literature."

Many manuals and guides have been written over the years to fill the training gap in scientific writing and speaking. Studying these (and I've been through a good many, more than once), I've found that they vary from the helpful and well-written to the dismal and discouraging. There are many excellent thoughts scattered through this literature, like glittering jewels in gray sand. But there is also much glass and cinder. What follows is a discussion of several core problems.

To begin with, a majority of books on scientific communication boil down to collections of rules, standards, and warnings. Some even claim to offer the opposite, but end up embracing the enemy. Such books will tell you: "Keep all your sentences short and simple," "Avoid emotional terms," "Employ the active tense whenever possible," "Nouns and verbs are the most important words in science," "Follow the IMRAD structure (Introduction, Methods, Results, and Discussion) in all your papers," and so on. This type of advice, if viewed with the rigor of its own prescriptions, becomes a list of absolutes, like Martin Luther's *Ninety-Five Theses,* to be nailed to the door of every science department in the land.

From a certain point of view, the learning of rules makes sense. Science, after all, is awash in protocols, principles, and standards. Why not apply this to writing? Certainly it can be done. But let us be clear about what it means. The real focus is less on learning per se than on obeying codes of authorial behavior. One is not encouraged to find examples of good scientific expression and to study them, but instead to follow a set of fixed and impersonal regulations. That is why these manuals so often adopt a tone of law enforcement. Their authors would like to use you, the writer in waiting, as an instrument of *habeas scriptus* regarding proper scientific English. That they have done so as part of a venerable tradition in American letters—what H. L. Mencken once termed "the melancholy order of the schoolmarm"—does not make their case any more attractive. Rule-driven advice can overwhelm a reader and validate any discomfort she or he already feels toward writing. Tiptoeing through a minefield of potential errors does little to advance confident steps toward the authorial act. Such advice thus tends to provide us more with the measure of our failures than aids to our success.

Let me give a specific example. Many manuals spend much space laying out precise standards for various items—references, tables, format, article structure, and so forth. All of this could more easily and usefully be learned from study of the existing literature in one's own field. This is especially true for journals in which you intend (or hope) to publish, since each tends to have its own specific requirements in any case. Actually, studying the literature is likely to be the *only* way to gain truly practical knowledge of this kind, as there really are no set conventions for science as a whole.

This brings up another problem area. Authors of writing guides in science tend to offer counsel that reflects their own (inevitably limited) experience. What is good for biomedicine or agronomy, however, is not necessarily good for chemistry or zoology. The supposed universal IMRAD structure is rarely, if ever, followed in large portions of the geosciences, mathematics, physics, engineering, and many other domains where fieldwork, theory, and descriptive efforts are on exhibit. Attempts to impose standards for all scientific papers, presentations, posters, and theses are therefore doomed to several degrees of irrelevance if performed on a detailed level. Instead of such hand-holding, advice is needed on *how to learn* what is acceptable and how to improve on it.

Such advice begins with an appreciation for the history, evolution, and current nature of technical discourse. Without this understanding, after all, ideas on how to create good scientific expression remain in something of a vacuum. Researchers tend to think of their discourse as eminently simple, almost pure in substance. Nothing could be farther from the truth. No less than in the humanities, or in any other form of discourse, scientific expression is full of strategies whose goal it is to persuade the wary or unwary reader. Think, for example, of the importance of logic and organization in our papers, how we use citations, the places where we feel it is appropriate or inappropriate to generalize, the ways in which we use illustrations. All these aspects are part of the rhetoric of science. Being conscious of how they work can only help us communicate better. Otherwise, as the chemists of our creations, we are only playing with pretty bottles and colorful liquids.

These criticisms being given, I feel the need to reverse direction. Having buried the competition, I now come to praise it. This is not disingenuous. It would be foolish, and indeed anti-intellectual and very unscientific, to claim that "the existing literature" can be wholly avoided or dismissed. On the contrary, my own perception is that a large volume of useful insight exists in this literature (even if it is sometimes compromised by the difficulties noted). Too many intelligent, well-meaning, and experienced individuals have written on the topic for this not to be the case. Moreover, though it may seem unwise of me as an author to say so, no single work is likely to be comprehensive enough to cover all aspects of scientific discourse adequately. In particular, I have found things to admire in such works as James Paradis and Muriel Zimmerman's *MIT Guide to Science and Engineering Communication,* Robert Day's *How to Write and Publish a Scientific Paper,* Vernon Booth's *Communicating in Science,* Edward Huth's *How to Write and Publish Papers in the Medical Sciences,* and David Porush's *Short Guide to Writing about Science.* If for no other reason, I would encourage you to read these and other works to test the validity of what I have said above and what will be presented below. There is also a good chance that I will meet some of these authors at a conference one day . . .

THE APPROACH OF THIS GUIDE

This is a book about professional scientific communication—what it is, where it has come from, how it can be achieved, understood, and improved. Though the focus is on written expression, there are also chapters on professional speaking, the Internet, dealing with the press, and other topics. If it is therefore mainly a book for scientists who write for other scientists, it has not left out the larger world of communication as a whole.

A fair bit of focus is given the most prevalent kind of scientific publication now in existence—the journal article. Though admittedly a small subset of the total range in technical expression, the journal article is obviously where the dominant corpus of science has come to reside. This may change in the future; new forms of exchange may emerge. But for now and the foreseeable future, both hard-copy and online science will continue to employ the well-proven forms of raw data and article-type publication. Writing professionally for colleagues will not disappear or diminish any time soon.

This is a book of advice, not rules; guidance, not demands. It is my experience, from years of publication in both the sciences and humanities, that scholars of any stripe learn best how to write well if they are addressed *as writers,* not as mere laborers, toiling in the mills and quarries of the word.

What does this mean? A certain shift in dignity, to begin with. But more to the point, it means providing you, the writer, with certain understanding, techniques, and attitudes that will aid you in gaining command over the language you produce and consume for a living. This I hope to do in three fundamental ways. First, I review some points on the nature and history of scientific discourse—this gives us context and a realistic sense of what we can expect of ourselves. Second, I maintain that good writing very often has a base in reading—I mean, reading as writers do, with a critical eye and an ear for quality, for what is worthy of imitation. This leads directly to the third and final point: good communicators learn from others, by identifying and studying examples of successful expression in their chosen field.

This last idea is probably the most important of all. It is a very old and deeply tested truth: authors acquire a comfort and facility for writing by first emulating the excellent work of others. This has always been true, and often admitted, for poets, novelists, playwrights, essayists, and scholars generally. Indeed, the use of models was a central aspect of Western education from at least the time of Quintilian (first century B.C.) down to the 1800s (why this changed is a complicated story). As a general method, it remains very much alive in the arts and humanities today. Experience teaches that scientists most often learn to write this way, too, though on a haphazard basis, since we don't tend to acknowledge it very much or make it an overt part of training.

The thoughtful use of models, however, has another prominent advantage. It allows you, the writer, to chose your own teachers (or coaches, if you prefer). Writing is a personal activity, as I have said. This complements the fact that being "authors," in any area, means we are part of a community of producers and can improve our skills by drawing on the good work that other members have done. I will have more to say on this matter below. For now, let me leave you with a famous phrase by one of America's preeminent poets, T. S. Eliot, who once, perhaps in a moment of confessional weakness, said: "It is only immature poets who borrow. Mature poets steal."

A word concerning what this book is *not* about. It is not about teaching you grammatical rules or proper scientific usage. There are other volumes along these lines;[1] this book assumes that, if English is your mother tongue, or a strong second language, you are able to form a competent sentence in this language, at least some of the time, and that

1. See, for example, Day 1995. In addition, a large number of other manuals on technical writing contain sections dealing with grammar and usage.

you know how to use a dictionary. If so, read on; this book is for you. To those with English as a second (or third or fourth) language, I have devoted a separate chapter. Those unable to write grammatically in any language should have been either bored or scared off by what has been said to this point.

Please think of using this book in several different ways. Chapters 2–4 form a unit, lay out the major themes, and will be most rewarding if read together (in order, if possible). Chapter 5 takes some of these themes to a higher level and may not be for everyone. Succeeding chapters, on the other hand, can be either perused one after another, or dipped into, one at a time, as need or interest arises. If nothing else, I would like you to come away from this book with a changed view of scientific expression—what it is, what makes it up, where it is going, and, above all, how to use it. If even part of this is achieved, a good deed will have been done.

A FINAL INTRODUCTORY WORD: PHILOSOPHIES OF LANGUAGE

I have said that the way in which one views language has an important effect on how one uses it. Scientists have been prey, for some time, to a particular philosophy of language that tends to derail their understanding of what might be termed "the scientific message." I refer, specifically, to the overriding maxim "Simplify, simplify." There are many variations on this theme; no doubt you've heard some of them: "Use as few words as possible," "Eliminate anything that is not essential," "Scientific writing must be transparent, a mere vehicle," "Use the active voice at all times," and so forth.

All such ideas exhibit a deep misconception about the nature of technical discourse. The "keep things simple" mentality is a way of declaring martial law on the inevitable complexities of scientific communication. Besides embodying a philosophy of distrust, this way of thinking lacks any appreciation for the rhetorical *flexibility* of technical writing, as a form of human expression, and the range of literary techniques such writing normally includes—indeed, *must* include. To persuade and convince a highly critical audience, authors cannot simply brain-dump information onto paper. If they could, there would certainly be no need for a book of this type. We would all be masters, with no need of apprenticeship.

Let me give an example. One of the rules most common to the "simplify" philosophy is that the scientific writer should do away with any and all phrases such as "under these or similar circumstances," "it is important to note," "for the most part," "it is doubtful that," and so on. These kinds of fragments, however, though perhaps inessential as far as the data goes, perform a required function in good writing. They act as transitions between sentences or paragraphs and also serve as helpful cues for the reader. They add pacing, flow, and important internal connection to the argument.

Effective arguments in any area of study, that is, employ a host of persuasive techniques. Many such techniques, in fact, are used equally by scientific and literary writing, though in different ways. This can be easily shown by a close analysis of any technical paper (see chapter 2). At a fundamental level, there is no deep divide between the sciences and the humanities when it comes to the basics of expression. Only, perhaps, a series of guarded trenches.

A main goal of this book is to help make scientific writers and speakers aware of the forms that they are using, or might use, when they produce competent science. This means learning to read with a critical eye and understanding how specialized the scientific message really is. Writing, in particular, is a messy business. It is as full of trial and error, dead ends, frustrated effort, and minor triumphs as any other part of research. What eventually emerges (hopefully) is a reasonably well organized, logical flow that hides this struggle. In the words of Peter Medawar, Nobel laureate in medicine and frequent author on matters of science, "the scientific paper is a fraud." But then, so is all successful writing.

Scientific communication is highly stylized—far more stylized, in fact, than forms such as the literary essay. When we look back at the past, say to the 17th century, and trace technical expression forward, we find that what we are doing when we write is telling very condensed, extremely formalized "stories" to an equally particular audience. In most cases, we have learned to do this through imitation, another trial-and-error process. Consciously or otherwise (usually otherwise), we are employing strategies to convince the reader of our knowledge, competence, originality, and contribution. This seems a tall order, when put this way. It is both ordinary and magnificent. Perhaps the vague sense that this is going on helps make us the critical, scrutinizing, and often skeptical beings that we are. But it should also reconnect us with the reasons why we originally chose to do science, the wonder and fascination, the ambitions and desires, that propelled us in this direction. Writing is about these aspects of our lives, too. Scientists are also writers because science is a great presence in the world.

CHAPTER TWO

Scientific Communication

Historical Realities for Readers and Writers

But that wasn't all. Control also had to be exercised during the actual writing process (which took about fifteen minutes per novel), and to do this the author had to sit, as it were, in the driver's seat, and pull a battery of labeled stops . . . By so doing, he was able continually to modulate or merge fifty different and variable qualities, such as tension, surprise, humor, pathos, and mystery. Numerous dials and gauges on the dashboard itself told him throughout exactly how far along he was with his work.

—Roald Dahl

MATTERS OF HISTORY

As scientists, we are largely creatures of the contemporary. Unlike other areas of study, our training provides us little on the history of our discipline or the development of our discourse. Does scientific language even *have* a real history? A legitimate question. But then, what would Newton or Laplace make of a recent article in *Physical Review Letters*? If Darwin's *Origin of Species* were submitted to a major scientific publisher today, what might be its chances of acceptance (should we say, survival)?

These are not merely academic questions. Language stands still for no one, and scientific discourse, as a subset of language generally, is no exception. If you doubt this, I urge you to read through articles in your own field written 50, 75, and 100 years ago or more: you'll find both obvious and subtle differences from the speech of today, and not only in terminology. Let me offer a few examples here.

> Anno: 1672. In the year 1666 . . . I procured me a triangular glass prism, to try therewith the celebrated phaenomena of colours . . . It was at first a very pleasing divertissement, to view the vivid and intense colours produced thereby; but after a while . . . I became surprised to see them in an oblong form; which, according to the received laws of refraction, I expected should have been circular. (Newton, "A New Theory about Light and Colours" [1958], acknowledged to be the first modern scientific paper)

> 1760. Beside the horizontal division of the earth into strata, these strata are again divided and shattered by many perpendicular fissures, which are in some places few and narrow, but oftentimes many and of considerable width. There are also many instances, where a particular stratum shall have almost no fissures at all, though the strata both above and below it are considerably broken: this happens frequently in clay, probably on account of the softness of it. (Mitchell 1970, 84)

> 1868. The extent to which a country suffers denudation at the present time is to be measured by the amount of mineral matter removed from its surface and carried into the sea. An attentive examination of this subject is calculated to throw some light on the vexed question of the origin of valleys and also on the value of geological time. (Geikie 1970, 523)

> 1965. In arid climates the rocks exposed to the blazing sun become intensely heated, and in consequence a thin outer shell expands and tends to pull away from the cooler layer a few centimeters within. Under perfectly dry conditions, however, the stresses so developed are insufficient to fracture fresh, massive rocks. Experiments leave no doubt about this. (Holmes 1965, 248)

> 1990. Regional patterns of present-day tectonic stress can be used to evaluate the forces acting on the lithosphere and to investigate intraplate seismicity. Most intraplate regions are characterized by a compressive stress regime; extension is limited almost entirely to thermally uplifted regimes. In several plates, the maximum horizontal stress is subparallel to the direction of absolute plate motion. (Zoback et al. 1990, 291)

Even in so brief a selection, we can see some of the crucial changes occurring in scientific expression over the span of the modern era. As early as the second excerpt, for example, the object of interest ("strata"), not the observer, now performs most of the action. Phenomena take over as the main characters of the "story." Because of this, the writing gains a more objective, formal tone. The confessional person, the

Newtonian "I," has been replaced by an observing god. But the last few examples reveal a still greater shift, a potent increase in density and reliance on terminology. By the final excerpt, nearly all conversational touches are gone; any puffs of informal air have been forced out, such that the style becomes akin (we might say) to a form of technology. Such is the direction our discourse has taken, and continues to take today. How, then, did this process begin?

WHERE WE CAME FROM: THE BEGINNINGS OF MODERN SCIENTIFIC EXPRESSION

Modern scientific writing in English began in the 17th century, with authors such as Francis Bacon, Robert Boyle, and Isaac Newton. This period was characterized by intense debates over the nature of language generally. At issue was the presumed power of words to control knowledge, as Bacon put it, to "force and overrule the understanding, throw all into confusion, and lead men away into numberless empty controversies and idle fancies." Bacon was thus the first to claim revolt against Elizabethan styles of writing (which, of course, included Shakespearean drama); these, he said, pulled a veil between the intellect and the world. To advance knowledge, especially "the new experimental philosophy," there was needed a simple, direct, and unadorned form of speech. This would lift the veil and provide "an equal number of words as of things."

Bacon's followers took his ideas very much to heart and made them a philosophical nucleus for the new Royal Society of London, the first scientific society in the English-speaking world. How closely did these men adhere to Baconian principles? Thomas Sprat, in his *History of the Royal Society* (1667, 4), gives us some idea:

> Who can behold, without indignation, how many mists and uncertainties, these specious tropes and figures have brought on our knowledge . . . [We of the Society] have therefore been more rigorous in putting in execution the only remedy that can be found for this extravagance; and that has been a constant resolution, to reject all the amplifications, digressions, and swellings of style; to return back to the primitive purity and shortness . . . [to] a close, naked, natural way of speaking . . . to bring all things as near the mathematical plainness as they can; and preferring the language of artisans, countrymen, and merchants, before that of wits, or scholars.

Those of the Royal Society were never more flowery than when denouncing the bloom of the Elizabethans.

Yet a new style did emerge, by the end of the century. The society had established a journal (the *Transactions*), which mainly published lectures given during meetings. That the earliest scientific papers in English very often had to be read aloud in front of an audience did, eventually, impose certain changes in style and length. Newton's "New Theory about Light and Colours" (1672) helped set a standard. The paper was written as a letter to Henry Oldenburg, then president of the society, in a form to be read aloud to the membership. Newton's paper showed how effective it was to confine one's speech to a demonstration, a repeating in words of what was done in actions. Newton, meanwhile, had drawn on, and simplified, the writing of Robert Boyle, who, as it happens, may well have modeled his own discussions of chemical experiments on the essays of Montaigne.[2] Boyle appears to have adapted the Montaignean "essay of experience" to the telling of thoughts, procedures, and results of his experiments. Newton abbreviated the form to a sort of plot summary of events and findings. And so, in large part, the scientific article has remained. The witnessing "I" was thus science's first storyteller. It was a way to "prove" rhetorically that the work had actually been done.

What happened thereafter, during the next three centuries, is a complex tale in itself. Different fields evolved somewhat separately, while sharing the article format and an overriding idea of what "scientific style" should be (Montgomery 1996, chap. 2). Yet literary elegance clearly had a place in science as recently as the end of the 19th century. Note, for example, a passage from the famous, ether-destroying paper by Michelson and Morley published in 1887:

> If the earth were a transparent body, it might perhaps be conceded, in view of the experiments just cited, that the inter-molecular aether was at rest in space, notwithstanding the motion of the earth in its orbit; but we have no right to extend the conclusion from these experiments to opaque bodies . . . [And] as Lorentz aptly remarks: "Quoi qu'il en soit, on fera bien, à mon avis, de ne pas se laisser guider, dans une question aussi importante, par des considérations sur le degré de probabilitié ou de simplicité de l'une ou de l'autre hypothèse." (334)[3]

This, indeed, seems a long way from the likes of today's article on superstring theory or quantum chromodynamics. When was the last time

2. These connections are discussed in Paradis 1987.

3. The French translates as, "Whatever the case, with respect to a question of such importance, one would do well in my opinion not to be swayed by considerations regarding the degree of probability or simplicity of one or another hypothesis."

you read (or wrote) a paper stating "we have no right to extend the conclusion" or quoting French? What would a contemporary editor do to such a passage?

Yet there is much else that has remained in place. Don't we still propose hypotheses in order to confirm or destroy them? Don't we cite the competition, or our immediate predecessors, in a manner that supports our approach and conclusions? Of course we do, though in a more formal style. What, then, of the Newtonian "I" and its fate over time? Was it really killed off, forced into extinction by a more objective style? In reality, no. Both rhetorical approaches have existed side-by-side, and even together, down through the centuries, up to the present, though, again, in stylized form:

> Anno: 1775. I cannot, at this distance of time, recollect what it was that I had in view in making this experiment; but I know I had no expectation of the real issue of it . . . If, however, I had not happened . . . to have had a lighted candle before me, I should probably never have made the trial. (Priestley 1952, 120)

> 1903. The results of the investigation of radio-active minerals . . . led M. Curie and myself to endeavour to extract a new radio-active body from pitchblende. Our method of procedure could only be based on radioactivity, as we know of no other property of the hypothetical substance. The following is the method pursued for a research based on this property. (Curie 1952, 522)

> 1999. We first searched for neurons exhibiting a relatively high rate of spontaneous activity when the animal's eyes were closed. Next we characterized the orientation tuning properties of these neurons and selected the neurons with sharp tuning preference and robust response. We chose orientation tuning . . . because the majority of neurons in cat striate cortex are tuned for the orientation of bars or gratings. (Tsodyks et al. 1999, 1722)

If the confessional "I" has turned into the royal scientific "we," the first person point of view is still an important element in our efforts at persuasion. We are still wedded to telling "stories." Yes, our language has tended to exchange tasteful tweed for gray flannel. Yes, we no longer write for someone who might be interested in an artful, novelistic type of narrative. But note how Priestley's confession of serendipity ("If I had not happened to have had a lighted candle") changes for the Curies, who are "led" by "results" to perform their experiments, and how, in the final example, "neurons" are the principal performers within a symphony of choices conducted by the "we." The tales we tell are, by nature, still based on narrative techniques whose goal it is to gain agreement and cooperation.

ROLE OF EDUCATION

Scientific language, therefore, does have a history. Indeed, our language continues to evolve, and will keep on doing so. Changes in technical expression over time are not due, as is so often believed, to armies of editors, working diligently to tame an ever more rebellious mob of error-prone professionals. Language evolves because of a host of factors, not all of them well identified, and very few of them planned. As the examples above suggest, individual scientists, editors, publishers, and institutions all have played and continue to play a role. Manuals on writing and usage do, too; these have very often been attempts to bring down the gavel and compel order—or stasis—on the shifting landscape beneath all our feet. The task, in some sense, appears at once heroic and impossible. In another sense, however, it is merely part of the process.

Until quite recently, roughly the last 75 years or so, modern scientists were educated to acquire the skills of good research *and* good writing. In the 18th and 19th centuries especially, scientists were regularly among the most eloquent authors of the day (one thinks of Sir Charles Lyell or Thomas Huxley in England, for example). This was due, in part, to the type of classical education then in effect, whereby all students at the middle and upper levels studied authors of the Greco-Latin tradition, as well as grammar and rhetoric, and the works of the most successful modern thinkers and writers as well. Learning to write in a variety of styles was part of this education.

During the past century, such training has given way to one that is far more specialist in design, far less interested in language as a medium. Again, the reasons for such a change are many and cannot be explained by a few pretty thoughts or ugly phrases (e.g. "Little science has become Big science"). The background required to train in a research discipline has broadened and deepened enormously. Learning complex methods and acquiring a huge technical vocabulary (a form of language learning) are but two of many requirements not faced by our predecessors.

We can think of it another way. The amount of research performed and published today is many orders of magnitude greater than even 80 years ago (for most of the 20th century, the volume of scientific literature doubled every 15 years; see Price 1963). As this has happened, the length of the average article has tended to shrink, and its style has gained the density of a rare-earth metal: think of the papers published in *Science* or *Nature*. These two journals, the most international in all of science, provide wonderful examples of what has happened to technical literature. In one way, these periodicals are holdovers from the past—they publish news and research from a wide array of disciplines. Yet in every other sense, they are leaders in specialization. Their papers are commonly less than five pages long and very difficult to understand (even when well written) except to practitioners in their respective, individual fields. In style, they frequently read like abstracts, and in content they are often models of frontline, influential work.

Why is any of this important? There are several reasons. First, the past does offer us both perspective and guidance. It confirms that scientific writers have succeeded by studying and imitating models of their craft. Using examples of good writing as analogs for one's own work has been a powerful and highly profitable "scientific method" throughout the modern era. Second, a sense of history reveals that the real job of a guide like this one is itself historical, that is, to help put back some of the awareness of language use that was formerly part of the scientist's training. Third, history shows that the scientist writes and speaks in the stylistic idiom of his or her time; this is an inescapable fact. As a participant in a technical field, you enter the flow of its evolution and add your contributions to the development of its discourse. Obviously, you don't have to understand this to be an active, competent scientist. But it gives you a decided advantage. It enhances your power of choice by offering a base to improve your communication skills over time, through self-directed effort, instead of betting against chance and looking for figures in the carpet.

SCIENTIFIC RHETORIC: AN INSTRUCTIVE ANALYSIS OF A NOTABLE PAPER

For the great majority of the modern era, science and literature openly employed many of the same techniques to persuade their readers. What about today? To answer this, I'd like to take a brief look at a fairly recent scientific paper of some renown, peel back the patina of discovery, and point up some of the rhetoric it employs. Here are the first few sections.

We wish to suggest a structure for the salt of deoxyribose nucleic acid (D.N.A.). This structure has novel features which are of considerable biological interest.

A structure for nucleic acid has already been proposed by Pauling and Corey. They kindly made their manuscript available to us in advance of publication. Their model consists of three intertwined chains, with the phosphates near the fibre axis, and the bases on the outside. In our opinion, this structure is unsatisfactory for two reasons: (1) We believe that the material which gives the X-ray diagrams is the salt, not the free acid. Without the acidic hydrogen atoms it is not clear what forces would hold the structure together, especially as the negatively charged phosphates near the axis will repel each other. (2) Some of the van der Waals distances appear to be too small.

Another three-chain structure has also been suggested by Fraser (in the press). In his model the phosphates are on the outside and the bases on the inside, linked together by hydrogen bonds. This structure as described is rather ill-defined, and for this reason we shall not comment on it.

We wish to put forward a radically different structure for the salt of deoxyribose nucleic acid. This structure has two helical chains each coiled round the same axis (see diagram). We have made the usual chemical assumptions, namely that each chain consists of phosphate di-ester groups joining β-D-deoxyribofuranose residues with 3′,5′ linkages. The two chains (but not their bases) are related by a dyad perpendicular to the fibre axis. Both chains follow right-handed helices, but owing to the dyad the sequences of the atoms in the two chains run in opposite directions. Each chain loosely resembles Furberg's model No. 1; that is, the bases are on the inside of the helix and the phosphates on the outside . . .

If it is assumed that the bases only occur in the structure in the most plausible tautomeric forms (that is, with the keto rather than the enol configurations) it is found that only specific pairs of bases can bond together. These pairs are: adenine (purine) with thymine (pyrimidine), and guanine (purine) with cytosine (pyrimidine).

In other words, if an adenine forms one member of a pair, on either chain, then on these assumptions the other member must be thymine; similarly for guanine and cytosine. The sequence of bases on a single chain does not appear to be restricted in any way. However, if only specific pairs of bases can be formed, it follows that if the sequence of bases on one chain is given, then the sequence on the other chain is automatically determined . . .

The previously published X-ray data on deoxyribose nucleic acid are

> insufficient for a rigorous test of our structure. So far as we can tell, it is roughly compatible with the experimental data, but it must be regarded as unproved until it has been checked against more exact results. Some of these are given in the following communications . . .
>
> It has not escaped our notice that the specific pairing we have postulated immediately suggests a possible copying mechanism for the genetic material. (Watson and Crick 1953, 737–738)

One of the most epochal papers in all of 20th-century science, Watson and Crick's article defies nearly every major rule you are likely to find in manuals on scientific writing. It does not follow the IMRAD (introduction, methods, results, and discussion) structure. It is entirely descriptive and not in the least analytical. It contains many "unnecessary" phrases ("we wish to," "in other words," "so far as we can tell"), vague words ("ill-defined," "loosely," "roughly compatible"), and even expressions of emotion ("of considerable interest," "kindly," "radically different"). The article has redundancies (the first sentence is repeated at the beginning of the fourth paragraph) and even grammatical errors (frequent use of the unrestricted "which" instead of "that"). Its paragraph form is uneven and improper. And finally (but this is far from a complete list), it has no real conclusion.

Yet, despite these unforgivable blunders, the article is convincing and effective to read. Why? Because of the organization and flow of the argument and how these are established. Let me discuss a few important elements. First, the style is highly concise. This, in fact, matches the main effort of the paper, which is to "build" a structure through specific details. The use of brevity, moreover, gives the impression of control, as if the authors were very carefully and consciously reporting only the most crucial portions of their work. We, as readers, are therefore treated as if we deserved to see only the best, most privileged information; we are given the authors' full confidence.

Second, there is the frequent use of "we," the testimonial technique we noted previously. This provides an immediate human presence, allowing for constant use of active voice. It also gives the impression that the authors are telling us their actual thought processes (in reality, of course, these were very messy; what we get here, as anywhere in science, is a strategically refined lie). Constant repetition of "we" and "our" tells us to associate this knowledge directly with the authors, that it is "theirs" in a sense, and no one else's. This may seem greedy or antiscientific. Aren't we supposed to be offering our knowledge to the world? Well, yes, of course, but publication is also about claiming credit for what we've done. Ideals aside, contemporary science is extremely competitive. Watson and Crick's use of the rhetorical "we" expresses this directly, by raising a flag on DNA. The rest, as they say, is history.

Third, the flow of the argument moves in a particular way: announcement (paragraph 1), dismissal of competitors (paragraphs 2, 3), main proposal (paragraphs 4, 5, 6), qualification (paragraph 7), and explosive finale (paragraph 8). What type of "story" does this offer? A tale full of understated sound and fury. The authors tell us that there is a rush to unravel the "grail" of DNA, that all important alternatives appear doomed, and that they (the authors) have happened upon the rightful path. Watson and Crick then reveal their discovery, build a castle out of detailed descriptions, set their flags flying, and note that this fortress has but a temporary weakness. A final fanfare confirms that the "grail" is indeed in their hands and that they will use it to effective ends. This last rhetorical flourish ("It has not escaped our notice") states, in other words, that there will be more, equally significant articles to come. Like a well-crafted story, the argument follows an hourglass structure overall: it begins with the general, moves into the ever more specific, and then, at the end, expands outward into the even more general.

Finally, the paper employs both enumeration and speculation. Most manuals of writing will tell you that a numbered list needs three or more items in it, and generally speaking, this is a fairly good rule to follow. Here, in the second paragraph, we see only two. Why? The authors are delivering summary judgment; they wish to eliminate their closest and most fierce competitor. Enumeration serves to take separate sword blows at the Pauling structure. First the head ("not clear what forces would hold the structure together"), then the limbs.

Speculation appears in the "if it is assumed . . . it is found . . ." structure used in paragraphs 5 and 6. This structure is among the most venerable in the history of Western rhetoric; volumes have been written on it. When looked at rigorously, it falls apart instantly, since "if" cannot possibly lead to "it is found"; assumptions alone are insufficient to support concrete discoveries. Yet the technique is effective and is found everywhere in science today. It has certain logical-emotional satisfactions that are persuasive if used well and in qualified fashion. By extending a known idea in unexpected directions, we can add a note of momentary drama (subtle to be sure), or make a relieving shift in an otherwise plodding course of explanation. The "if . . . then" technique is part of the rhetorical toolbox of the competent scientific writer.

Looked at in this fashion, a technical article may seem like a thinly veiled series of indiscretions. Yet such is the nature of writing to persuade: whether with intent or intuition, we employ strategies to gain the cooperation of our readers. Indeed, any scientific text can be analyzed in this manner. A full rhetorical study of the Watson and Crick paper could fill a monograph, and would reveal techniques (including the above) that frequently occur in literary writing, too. Such an analysis would not be at all excessive or without value. Good writing uses organic "technique" to guide the experience of the reader. When we look beneath the formality of scientific writing, today or in the past, we can see this happening. It is like gazing into a drop of pond water through a microscope.

All of this begs a question: are good writers aware of the techniques they employ? The answer is almost always yes and no. Yes, because good writers very often plot out or experiment with the logical course of their narrative. No, because many specific rhetorical techniques are used intuitively, even in standard fashion; they have been learned by attentive reading of the literature and imitation of it. Effective writers are those who have an inner ear for what sounds right, what is persuasive at each turn of a discussion. Being aware of even a few such techniques and how to acquire them will provide the scientist with a powerful instrument for his or her expressive work.

GRAMMAR: FACTS AND FALLACIES

Questions of history and language change bring us, happily or unhappily, to the subject of grammar. What is to be said about it? Look again at the first set of examples given in this chapter. You will see a kind of equal and opposite reaction between two elements: as the number of technical terms has grown, the grammar of the sentences has simplified. This is not an accident. Consider the following:

> A comprehensive overview of quality control in DNA would include a discussion of DNA polymerase fidelity and postreplicative mismatch correction and would also consider the damage-responsive cell-cycle checkpoints and the signal transduction systems that lead to cellular effects. (Lindahl and Wood 1999, 1898)

Now replace each technical term or phrase with an ellipsis:

> A comprehensive overview of . . . would include a discussion of . . . and . . . and would also consider the . . . and the . . . that lead to cellular effects.

Finally, with a bit more distillation, we get

> A comprehensive overview of . . . would include a discussion of . . . and would also consider . . .

Once the terminological smoke clears, the average scientific sentence today emerges as fairly elementary. If you still doubt this, perform the same exercise on an entire paragraph in a recent article—dare I say, on your own writing? I guarantee it will be revealing. Indeed, the exercise can be valuable in pointing out unneeded complexity (the above sentence, for example, could certainly be helped by enumeration). But note: the mixture of complicated vocabulary and relatively simple grammar is a volatile one. It makes it easy for scientists to rely too heavily on terminology for meaning and ignore the need to craft good sentences. This is a point to remember.

I ask that you neither let the subject of grammar cripple you with concern nor float your intentions too high. Good grammar alone does not a writer make; bad grammar, if only occasional, does not destroy one. On the other hand, to communicate effectively, you have to be able to produce a decent sentence. This goes (almost) without saying. But being hyperconscious of possible mistakes can reduce your progress to a glacial melt. Perfectionism makes one unforgiving, both as a writer and reviewer.

The belief that scientists, in general, are unable (or unwilling) to construct a legible sentence has led a majority of writing guides and commentators to place considerable emphasis on "proper scientific English." Teaching grammar and usage, however, is a different job than teaching writing or speaking. Grammar is mainly about ingredients and formulae; it necessarily involves a rule-driven, mechanistic view of language use. Usage, on the other hand, is, in reality, a much more slippery subject and, perhaps because of this, tends to encourage police activity. Authors of technical writing manuals are constantly in riot gear over whether to change "prior to" to "before," "perform" to "do," and so forth. Such efforts are largely ineffectual and, worse, irrelevant. In many cases, they represent pet peeves. Indeed, different editors of different journals are likely to have their own, or to abide them to widely differing degrees, so learning them all is a distinct waste of time. The fact is that scientific discourse rumbles on, fortissimo, more productive than ever, without paying much attention to these constabulary proclamations.

Does this mean "standards" are out the window? Is science groaning under the weight of its own fatty, acidic verbiage? Hardly. Yes, there is poor writing in abundance (as in other fields); and indeed, it is important to do something about it. But implying that a few dozen ironclad rules will improve scientific expression overall is tantamount to believing that knowledge of the periodic table automatically makes one a chemist. If you think my characterization harsh, I suggest you read through Strunk and White's *Elements of Style* at a single sitting. Then, without the aid of a double martini, try to compose a paragraph or two.

Writing and speaking apply the local formulae of grammatical law to the open-ended, often chaotic flow of actual communication. Learning how to create, direct, and manipulate this flow requires an understanding (at some level) of how to organize meaning and give it a reasonably consistent logic. Using grammar to try to impose such logic is like using equations of fluid behavior to try to predict how a river system will evolve over millennia. Grammar is the zero law of the communication process: necessary but very far from sufficient.

Thus, my advice is this. If you feel shaky about your ability to write a correct sentence, to the point of near paralysis (or if editors tend to return your manuscripts stained with crocodile tears), then by all means focus your efforts first in this arena. Study a grammar text for a month or two, or more; gain a degree of confidence here. Learn the basics before proceeding to the delights, challenges, and deeper humiliations of professional writing.

THE ROLE OF SOFTWARE

Many readers will be aware that computer software now makes it possible to do a number of things automatically that used to be done by hand and head. This includes checking spelling, indexing, footnoting, outlining, creating tables of contents, and even summarizing documents (abstracts, executive summaries, etc.). Grammar, too, can now be checked and "corrected," even (presumably) tuned to formal, colloquial, technical, or custom settings. Some scientists I know find this useful, though selectively. Others simply hate it, both in practice and principle, since it is inflexible and therefore mechanical (which, of course, is actually what it is). There seems little doubt that software designers will continue to strive for advances in this area. An ultimate goal, I would guess, is for such a system to adapt itself to the style of the individual author. A noble and profitable aim, to be sure. But let us be clear about one thing.

These writing aids, taken together, betray an ancient and broken dream: namely, that technology can free us from our own, inevitably faulted, humanity. Useful aids, however, are not now, nor will they ever be, a substitute for applying careful thought to the process of writing and revising. The best editors will tell you: use these aids, by all means; experiment with them, so that you find how and where they can help you save time. But don't employ them as a technofix. Don't think they can do your writing for you, even locally. Overreliance on them has resulted in many a problematic manuscript and thus a waste of time for writer and editor both.

The one area of software development that might, before long, prove an aid to the actual writing process is that of voice recognition. For those few scientists able to acquire enough skill to dictate their articles (or parts thereof), this kind of program might well prove helpful, even a godsend, particularly if they can't type. Yet, again, the result is only as good as the original writer/speaker. Writing technology is a great advance, but it will not replace writing—unless, of course, we are to take seriously the success of Roald Dahl's hero in "The Great Automatic Grammatisator":

> And this was true, for within another couple of months, the genius of Adolph Knipe had not only adapted the machine for novel writing, but had constructed a marvelous new control system which enabled the author to pre-select literally any type of plot and any style of writing he desired ... First, by depressing one of a series of master buttons, the writer made his primary decision: historical, satirical, philosophical, political, romantic, erotic, humorous, and straight. Then, from the second row, he chose his theme: army life, pioneer days, civil war, world war, racial problem, wild west, country life, childhood memories . . . and many many more. The third row of buttons gave a choice of literary style: classical, whimsical, racy, Hemingway, Faulkner, Joyce, feminine, etc. . . . Finally, there was the question of "passion." [For this], he had devised an independent control consisting of two sensitive sliding adjustors operated by footpedals. (Dahl 1953, 57)

WHAT IT MEANS TO WRITE OR SPEAK WELL IN SCIENCE

Notwithstanding Dr. Knipe's achievement, when it comes to putting words on paper, the scientist has a certain advantage. What, possibly, could this be (don't all scientists *hate* to write)? Simply this: at a basic level, scientists have a choice generally denied to other disciplines. They can be purely functional writers—ordinary engineers of the word—or they can strive for higher levels of eloquence, even creativity, within the bounds of acceptable formality.

Outside of science, functional writing is commonly looked upon as dull and unskillful. In most fields, there are expectations (or hopes) of grace, color, style. Particularly in the arts, in history or literature, language is expected to call attention to itself, to at least strive now and then for some obvious sign of craft, cleverness, or felicitous phrasing. Even in business or sociology, material with the aesthetic quality of cement is viewed as lacking in something. Not so in science. We can be as flat and gray as we like and not be judged ill for it. Functional writers, in fact, make up the great majority of successfully published scientists. This does not mean that such writing is always good—but a significant portion *is* competent: readable, informative, and adequately organized. Moreover, few writers are bad (or good) all the time; varieties of competence tend to exist in any single piece of writing. Functional communication, at a proficient level, is very much something to strive for in science. Indeed, as we've already indicated, it is essential for good science to be done.

There is something else here, too. Nonfiction authors frequently come up against a number of fundamental questions. William Zinsser, for one, has laid these out nicely. He notes how they include (but are not limited to) the likes of, How am I going to address the reader? (Reporter? Provider of information? Average man or woman?) What pronoun and tense am I going to use? What style? (Impersonal reportorial? Personal but formal? Personal and casual?) What attitude am I going to take toward the material? (Involved? Detached? Judgmental? Ironic? Amused?) Who will my readers be? What sorts of publication venues might be interested in my work? How much will I get paid?

Nearly all such questions are irrelevant in science. We simply don't need to worry about them. They've already been answered, in large part, by history. Such is a benefit to professional discourse in almost any field, but much more so in science. Again, this does *not* mean that our speech is simple and unchallenging, not in the least. We have our own problems to solve that other nonfiction writers lack: how to translate data into words; how to describe experiments so that they might be repeated; how to use illustrations; which colleagues to cite or challenge.

But can we really choose to be eloquent writers in science? Assuredly we can. Here is where aspects like refined organization, use of transitions, sentence rhythm and length, and strategic employment of rhetorical technique come in. An entire chapter of this book is devoted to showing how some of these aspects can be used creatively. The trouble is, such creativity needs to be subtle in science. It reveals itself most often in an occasional manner and in background elements that direct and propel the argument, but quietly. It may therefore go largely unrecognized by a majority of readers. This is the risk you run in crafting beautiful science: only editors, writers such as yourself, and teachers of writing are likely to appreciate what you've done, at least at first. In the long run, of course, you are also in jeopardy of being used as a model for others.

I began this chapter with the contemporary scientific author, and it makes sense to end that way. History proves that, as writers and speakers, we are immediate contributors to the evolution of scientific discourse. Every article or proposal or report that we produce, every word we put to paper, is an event within the flow of this evolution. Scientific language continues to change, as it must. We are its creators and metabolizers. But even more, we are its primary medium: through our efforts to create and exchange knowledge, this language is made real and alive. To write and speak well, whether functionally or eloquently, is to take responsibility for history, for knowledge, for oneself as a scholar.

CHAPTER THREE

Reading Well

The First Step to Writing Well

The only demand I make of my reader is that he should devote his whole life to my works.

—James Joyce

THE CONCEPT OF AUTHORIAL EAR

Good musicians and skilled writers have something in common. They both have developed an ear for what sounds right and what does not. When faced with a work of music or text, the well-tuned ear can listen for the movement of notes and words, certainly, but it will be equally alert to patterns of sense, to elements of order and logic and how they move within the work. Having this type of skill is no small thing. But it is no mystery either. It involves being attentive to the medium in particular ways. It means, for a scientist-author, being able to detect what feels awkward in a sentence like this:

> In higher plants, flowering—the transition from vegetative to reproductive growth phase—is controlled via several interacting pathways influenced by both endogenous factors and environmental conditions.

This is the opening to a recently published paper in a prominent journal. It is not a wonderful sentence, but it is comprehensible. Suppose, however, it were written

> Flowering in higher plants is defined as the transition from vegetative to reproductive growth phase. This change is controlled by several interacting pathways,

> each of which is influenced both by endogenous factors and environmental conditions.

Breaking the original sentence up like this creates a very different sound and flow. The information becomes more pleasant to read, easier to remember. A short introductory definition of a fundamental process is followed by a longer sentence, explaining its controls. Sound and sense go together; one mirrors the other. Ideally, the next sentence would be a bit longer and would either explain the noted pathways or introduce a species of plant whose growth activity has been investigated within the context just presented.

Every piece of music and each scientific article is an attempt to transfer something to the audience, not merely to "express" or "publish" it. Writing, in particular, is always a form of teaching—an attempt to give readers what they did not have before. Being sensitive to this process counts as an invaluable advantage. Recognizing good writing for what it is can be the first step to actually doing it yourself.

INTERNALIZING PREFERENCES: THE VALUE OF MODELS

Is it possible for the "ordinary" scientist to acquire such an ear? Musicians, we know, have talent. But talent must be trained and developed and, in any case, can be partly defined as a heightened ability to imitate and go beyond provided material. What about writers? Certainly talent, though vaguely understood, may exist here too. But skilled writers in a great majority of cases—and especially in the professions—are made, not born. How? The most natural and effective process, and the one most often followed throughout history, is that of apprenticeship. Such, indeed, is really how any complex skill is acquired, sooner or later.

We first learn how to do research this way, by imitating teachers, older colleagues, and so on. Throughout our career, moreover, we are likely to absorb tips and techniques from our betters and those that help us in our work. Apprenticeship involves disciplined emulation; you gain ability over time by incorporating and adapting the good work of others, so that you eventually develop a style of your own. Even the most gifted musician moves through a succession of composers and pieces, writing compositions that are at first wholly imitative, often painfully so. A personal style emerges as the composer or writer internalizes and mutates what she or he has selected as most worthy or worthwhile.

A deep difference between science and music is that, for scientists, this type of learning is ordinarily scattered, sporadic, unsystematic. Researchers are often left to teach themselves, despite the benefit of one or two writing courses during college. Imagine if this were true for musicians, journalists, or historians. Yet the scientist, whose work is no less involved in putting ideas down on paper, is largely abandoned to his or her own devices here.

This being said, where does one begin? Lacking a mentor or journeyman on hand, we turn to their embodiment—examples of solid, successful writing. Apprenticeship here involves two main activities: first, collecting examples of especially good writing whenever you come across them, and second, going over them in an attentive manner, so that, sooner or later, they become internal guides for your own sense of what sounds good and what doesn't.

The main thrust is to identify models of effective writing, study them, and then find ways to emulate them. In choosing your examples, you might select entire articles or only parts thereof, whether individual sections, paragraphs, or illustrations. In studying what you have chosen, you might begin by trying to understand what it is that appeals to you: Is it the flow and rhythm of the sentences, clarity of the wording, an inventive argument, the visual organization of an illustration, nicely turned phrasing, all of the above? To emulate, meanwhile, you can reread your models, copy them out, memorize and recite them, write a paragraph in their style. Many techniques exist. But the final goal is the same: make these models your own, internalize them. This is how you can put them to work for you.

One final point. Manuals on technical writing, and editors too, have often emphasized how much bad writing there is in science. "All are agreed," said a one-time editor of *Science,* "that the articles in our journals—even the journals with the highest standards—are, by and large, poorly written" (Woodford 1967, 743). Actually, I am one who does not agree (and I know many who also do not). This statement is a drastic oversimplification and, in fact, something of an insult to the good writers who do exist. No need to justify a moral high ground for those who, in the midst of their fatalism, would have us all be scientific Shakespeares. Really good writing, admittedly, is rare in most fields of endeavor—but this doesn't mean it is absent, not in the least. Seek and ye shall find it.

BEING A CRITICAL READER

To identify models, we need to read others' work in a particular way. We need to be alert to rhythm and sense, as well as logic—all in all, how the "story" is told. To do this may take a bit of practice. Scientists are taught to look for "content" above all else; this is how we have been socialized to read. But every piece of writing has multiple levels of sense within it, as the examples in chapter 2 revealed. When we read a beginning like this,

> Galactic dark matter may consist of weakly interacting particles which can be captured and trapped in stars, and which would then contribute to the transfer of energy. A special class of these particles ("cosmions"), with weak cross-sections that are larger than standard has been invoked as a solution of the solar-neutrino problem, and also as a means of suppressing convection in the cores of horizontal-branch stars. (Dearborn et al. 1990, 347)

we know, if only at the back of our minds, that we are in for a bit of heavy weather. Too much is being packed into these sentences; the reader is being treated as a highly absorbent material. We find ourselves having to reread in order to get what we need. Now look at another article opening:

> Subduction of the Juan de Fuca and Gorda plates has presented earth scientists with a dilemma. Despite compelling evidence of active plate convergence, subduction on the Cascadia zone has often been viewed as a relatively benign tectonic process. There is no deep oceanic trench off the coast; there is no extensive Benioff-Wadati seismicity zone; and most puzzling of all, there have not been any historic low-angle thrust earthquakes between the continental and subducted plates. (Heaton and Hartzell 1986, 675)

A passage like this makes it clear that technical language can be a pleasure. Notice how each point or thought is clearly marked off from the others. The writing has pacing, flow, and elegance. There is a diversity in vocabulary ("dilemma," "compelling," "benign," "puzzling") that draws us in, emotionally and intellectually. The sentences set up an excellent rhythm (short, medium, and long with breaks) that moves us along, deeper into the material. And beyond all this, the words simply sound good to the ear. Everything about the writing makes us want to read further, with anticipation.

These are the sorts of aspects that we need to be sensitive to, as critical readers. There is more than this, however. If an article or paragraph seems well-written, like the above example, take a moment or two to look at how it is organized. In the last excerpt, for example, notice how the writing moves from the general to the specific, how it adds detail to the topic at issue. Notice too that what is being discussed is the *absence* of certain phenomena—an excellent way to establish the "problem."

This type of analysis is good for short sections that you find appealing. For whole articles that seem well-written, on the other hand, you might check the overall logic by reading through the abstract, then the headings, surveying the illustrations, looking at the conclusions, getting an idea of how well it all fits together. Very often, you will find a mixture of things: admirable passages separated by mediocre writing; good style compromised by hurried organization; poor text offset by high-quality illustrations. More rarely, you will discover an article, report, or proposal in which all elements work together in excellent fashion. Save it, study it, make it part of your repertoire.

A FEW ISSUES TO CONSIDER

Scientific articles and reports are commonly made up of different sections—abstract, introduction, methods, and so forth—many of which differ among fields. Each of these sections takes up a distinct type of content and, to a significant degree, uses a distinct style of writing, a different voice. A good introduction sounds different than a methods section, which in turn should be stylistically separate from a discussion of results. An article tells its story by assembling these different voices and weaving them together. Transitions are essential (look for them!), but seams remain. Recognizing such complexity is part of the work you should do in perusing the literature. It shouldn't dismay or overwhelm you in the least—after all, it is achieved every day by thousands of scientists, in every field.

Students and unpublished scientists may feel that nearly all material in the major journals is worthy. After all, it got published didn't it? But articles appear in journals for a number of reasons, and quality of writing (unfortunately enough) is often not among them. A fair (or unfair) amount of bad text gets a day in the sun. Editors have demanding jobs and cannot possibly rewrite every piece of poor writing on important topics. Thus, a large amount of flawed text finds its way into even the best journals. You might check the howlers provided by Booth (1993) and Day (1995). As they note, the most common slipups are the improper adjective, for example, "crustal geoscientists" or "infant experts" (i.e. cantankerous researchers versus babies with brains) and the infamous dangling modifier: "Having been placed in an oven at 575°, we dried the sample for 3 hours" (no comment, in memoriam).

I would suggest that you collect some of these, too, as negative models, useful reminders. Nothing succeeds like the threat of humiliation.

Critical reading therefore means making judgments, being judgmental. We are trained as scientists to be evaluative, even skeptical, of each others' work. We need to extend this into the realm of expression, too. Perhaps the most simple, and effective, way to initially judge a piece of writing is to ask yourself: is this something I wish I had done myself, or am I glad I didn't? Once you have answered this question, you can go on to analyze the reasons for your response.

SOME TECHNIQUES TO CONSIDER

The following methods are suggested for choosing and using models. Note that I am *not* advising you to use all these methods. Different techniques will be useful (or possible) for different people and situations. Use what follows according to your needs, available time, and inclination. You should feel free to modify any of them or even come up with your own.

Choosing Models

1. To gain a better feeling for the sound and rhythm of language, read high-quality older literature in your field, say from 100 years ago (e.g. Pasteur in biology, C. G. Gilbert in geology, Kelvin in physics, Gibbs in chemistry), and compare this with a contemporary article or book. This exercise will help tune your ear to differences in language flow. (Note: this is not intended to suggest older literature as a model of style for you, only as a counterpoint to help sensitize your ear.)

2. If you're unsure about how selective to be, begin by making a fair number of choices and then whittle the totality down, bit by bit, as you reread, to the very best. Over time, you'll find that certain selections will continue to impress, whereas others will lose their luster. If authors in your field are known to be good writers, you might check their work first. Don't, however, confuse abundant publication with good authorship. You are looking for examples of superior expression, not

impressive credentials. Whatever strikes you as particularly well done is worth collecting.

3. Note the authors of examples you have chosen; read other work by them to see if it is of equal quality. You may find that you are especially attracted to the writing of only a few authors. This is fine; it will help make your selection easier.

4. Think about forming a reading group or reading seminar with some of your colleagues. A brief (e.g. one-hour) session each week to discuss the current literature in your field would be a good basis for evaluating and selecting examples of good (and bad) writing. This is something that is done fairly often in the humanities but, in my experience at least, very rarely in the sciences.

5. Make sure that you know what venue your samples came from. Textbook writing is not the same as journal writing, which is not the same as proposal writing. Each has its own specific requirements for writing style and logic. Be alert to such differences.

6. If possible, eventually choose a small number of models (i.e. five to ten) that you find particularly admirable. Stick with these for a while, until your judgment matures. Then think about replacing some of them with newer examples when they come along. Most important of all is to choose examples that you wish you had written yourself.

Using Models

1. Reread your chosen models on a regular, or fairly regular, basis. Make it a habit to go over them, if only briefly, at particular times when you have a few minutes free.

2. Try reciting out loud or copying in longhand those passages that you especially admire. This is a time-honored technique that helps sharpen your awareness and makes each passage more immediate.

3. Choose one sample, say the first paragraph of an article introduction, read it through carefully several times, paying close attention to such things as sentence length, word choice, and use of punctuation. Write a paragraph that might come next in the article, imitating the same style (it doesn't matter whether the information is accurate or even real). If this is difficult, then copy the model paragraph directly, changing some of the words.

4. Try taking a document of your own and rewriting a paragraph or two in the style of your chosen sample (i.e. as you might expect your selected author to have done).

WRITER'S BLOCK: A DIFFERENT PERSPECTIVE

Writer's block is a subject on which innumerable authors have weighed in. Ideas about its nature and origin are legion; solutions, however, go wanting. In part, this is because writing—of any type—is a very personal affair, thus so are inhibitions and anxieties associated with it.

Here I define this state as one in which all forward progress in getting words on paper is halted. This might happen when one first sits down to write, when there is a struggle to find the right word, when earlier composed material seems hopelessly bad, or when the entire effort feels doomed. Different authors have developed different methods for dealing with such interruptions. Some methods are mechanical (e.g. "Leave a space for the missing word and go on," "Skip several lines and start on the next idea," etc.); others are ecological ("Clean and reorganize your office, take a walk in the country, go for a run, get married, then sit down again"). One well-regarded source (William Zinsser) dodges the problem in this way: "There are as many kinds of writer's block as there are writers . . . and my name isn't Sigmund Freud" (Zinsser 1985, 23).

I would like to propose something a little different. Writer's block, we might say, occurs when the authorial ear—that inner voice of ours that brings forth words—is stopped or jammed. Something is needed to help that voice begin speaking again. Let me suggest a few methods for doing so, along the lines of what I have been saying.

- Go back to your models and read through selected passages on topics as similar to your own as you can find. Recite or copy them out, if necessary.
- Read a past article of your own, preferably on a subject not too distant from the one you're working on.
- Discuss your article or research with a colleague. Use this as an opportunity to explain the particular point or section where you're stuck. Listen for useful phrases in your descriptions and explanations; jot them down.
- If writer's block occurs when your article is already partly written, read what you have thus far, from the beginning up to the ending point. This may help gain new momentum.

The basic idea is to find a way to restart the flow of language and confidence within the authorial ear. None of the above can be considered fail-safe. The notion of reciting out loud or copying in longhand someone else's work might seem childish or embarrassing; if so, don't do it, except perhaps as a final resort. A fair number of my students have found that simply reading through articles on similar topics helps defrost the fluid of language. Others prefer to get closer to the actual words by hearing themselves say them or write them out. In extreme cases, I have advised an author to find a paragraph or section very close to the one he or she is stuck on, in subject matter and intent, and to rewrite it, sentence by sentence, changing phrases and terms where necessary but leaving a fair bit of the original intact. This is where "the common language of science" (as Einstein called it) can come to the rescue.

T. S. ELIOT AND THE IMPORTANCE OF THEFT

As I noted in chapter 1, it was the famous poet T. S. Eliot who said that immature poets borrow; mature poets steal. The reason this statement is well-known among professional writers is that very nearly all recognize its truth, but few wish to admit it directly. Using others' work as a guide for one's own, however, is as old as authorship itself. Adopting tricks of the trade from colleagues is done everywhere else in research—why not in writing?

If you see an organizational scheme, a paragraph structure, a phrase that you especially like, make it your own. You might, for example, find the type of introduction given in the passage above on subduction of the Juan de Fuca and Gorda plates particularly effective (I do). If so, imitate it directly; add it to your toolbox. Here's an example to show what I mean:

> Model passage. Subduction of the Juan de Fuca and Gorda plates has presented earth scientists with a dilemma. Despite compelling evidence of active plate convergence, subduction on the Cascadia zone has often been viewed as a relatively benign tectonic process. There is no deep oceanic trench off the coast; there is no extensive Benioff-Wadati seismicity zone; and most puzzling of all, there have not been any historic low-angle thrust earthquakes between the continental and subducted plates. (Heaton and Hartzell 1986, 675)

> Adoption. The precise mechanisms by which granite is emplaced in continental crust have continued to elude researchers. Despite abundant evidence for intrusion along fault segments, emplacement has been typically viewed as a relatively passive process. Granites are commonly undeformed; they show few signs of internal lineation; and, above all, their

> contacts with surrounding rocks usually exhibit only small amounts of shearing.

Clearly, this is not plagiarism. It is the sort of thing that writers speak of when they mention they have been "influenced" by another author. Frankly, once you begin to read the literature with this kind of eye, looking for potential influences, you will see that *all* writers, even the most influential, constantly "borrow" and "steal" from each other. It is, in fact, the most natural thing in the world and is practiced in every form of human expression, including music and art. Much of it, to be sure, occurs unconsciously (this is what T. S. Eliot meant by "borrowing"). Why not make it conscious, therefore less haphazard?

No writer or musician or artist is a Galápagos. As scientists, especially, we inhabit a mainland of both private and communal expression. Imitation, leading to adaptation, is the sincerest form of advancing our survival in this setting.

CHAPTER FOUR

Writing Well

A Few Basics

FUNCTIONAL EXPRESSION

Scientists have two basic choices as communicators. They can aim to be proficient and functional, or they can strive for higher levels of literary skill, even mastery. The first of these choices is for everyone; the second is not. Room exists in science for both, and both certainly exist.

Functional communicators are able to write and speak accurately, with reasonable precision, in a clearly organized fashion, without too many significant grammatical and syntactic errors. This mode is preferred by a majority of professionals. If achieved, it is not only acceptable, but wholly admirable. Functional communication embodies the philosophy that writing and speaking are methods for making knowledge available in an efficient, usable manner. This philosophy is a good one—as long as it stays within its own, limited context (as a general outlook, however, it is likely to be stultifying).

To write proficiently, you need a number of skills. You need a sense of good and bad grammar, an ability to impose order (sometimes out of chaos), the power to think visually (for illustrations), an ear for monologue. You need something else, too, however.

Writing demands, absolutely requires, a type of intense concentration, like chess or playing an instrument. This, too, is a skill. Scientists already possess it, in trained form, with regard to performing research or fieldwork. The same type of consistent focus you devote to observing and recording must be transferred to writing. Generating new knowledge is a creative act, with two scenes—investigation and composition.

TO WRITE IS TO EXPERIMENT

Writing is a process of experimentation. This is a crucial reality for scientists, and indeed for all professionals. Producing good, functional documents involves trying things out, engaging in trial and error, tinkering around. Even if you have a very firm and clear idea of the text you want to write, what finally emerges will rarely accord with this image in any precise way. There are simply too many elements that need to be worked out, too many levels of detail and decision making. And each of us is too much the human being, at once faulted and engaged, to act like a machine in this capacity. All of which helps us understand why writing is such hard work (for it is).

It can be an enormous help to know from the very beginning that you'll be entering this trial-and-error process. Experienced writers anticipate that there will be dead ends, struggles, and triumphs in lowercase. They know that they may even stumble across new, unforeseen ideas, those that illuminate their current project in a different way or that open a prospect for future work. Writing often involves such discovery—this, too, is part of its experimental nature.

Experimentation may continue up to the time that your document leaves your hands and goes out for review or publication. At this point, it is no longer yours: it belongs to the world. The document will now speak for itself. There is no "what I wanted to say" or "they'll know what I mean." There are only the words on the page as you left them.

I ask, finally, that you avoid one error of belief that is monstrously prevalent. This is the widespread notion that "to write clearly, you must first think clearly." This sharp little maxim may appear logical, but it is really rubbish. No matter how rational your thought may be (or appear to be) on a particular problem, no matter how detailed your intentions and plottings, the act of writing will almost always prove rebellious, full of unforeseen difficulties, sidetracks, blind alleys, revelations. Good, clear writing—writing that teaches and informs without confusion—emerges from a process of struggle, or if you prefer, litigation. This is true irrespective of how experienced an author may be, how many dozens or hundreds of papers she or he may have published. Most often, the terms of the formula given above need to be reversed: "clear thinking can emerge from clear writing." Imposing order by organizing and expressing ideas has great power to clarify. In many cases, writing is the process through which scientists come to understand the real form and implications of their work. But you must always be prepared for the process to be messy, experimental. If it were not, the best advice any guide could offer would be to see a therapist.

THE READER

Every piece of writing has an intended reader. This reader in science is likely to be a colleague, an interested peer who hasn't yet learned of your work. If your writing is effective, this reader will be fairly consistent.

The first few lines of a document establish the reader. An article that begins

> The nicotine acetylcholine receptor is a ligand-gated channel that mediates signaling at the vertebrate neuromuscular junction.

immediately weeds out anyone unfamiliar with the terminology given. The reader here is obviously a specialist. Starting right in with detailed subject matter like this, though sometimes condemned, is fully accepted practice in many areas of science. Certainly, it makes sense from the point of view of efficiency.

On the other hand, a beginning like this

> In recent years, there has been a marked resurgence of interest in artificial neural networks.

or this

> We report here results from the first stellar occultation by Saturn's giant moon, Titan, ever observed.

seems written for a more general reader. In each case, the second line might be crucial:

> Such networks consist of computer systems designed to imitate, in streamlined form, certain basic principles of operation of the human brain.

> Data obtained include the occultation chord at each station listed in Table 1, using half-intensity times, *t*, when the fraction of unocculted stellar light hitting the Earth was 0.5.

If, in the first example, the readers remain general, in the second they are suddenly whittled down to professional and high-level amateur astronomers.

Both of these approaches, too, are common practice. If the first is commendable for consistency, the second begins with a bit of flourish to stimulate interest, and then gets down to work right away.

The important thing is to install your true reader in the first several lines and then stay the course. Many writers make the mistake of dragging out their general openings, and then suddenly shifting to specialist discourse, or jumping back and forth between the general and specialist reader. This can frustrate and confuse the audience.

Check some of your models; see how they begin. Look at the first one or two paragraphs and see who the ideal readers are, how they are set up. Examine different journals in your field; get an idea of what is common, what works, and what doesn't. Reading for "the reader in the text" offers a valuable tool to evaluate your own and others' work.

THE AUTHOR

If every document contains a reader, it also speaks to this reader with a particular voice. In science, this is necessarily a voice of authority, a giver of knowledge. Let's look again at one of the examples above:

> In recent years, there has been a marked resurgence of interest in artificial neural networks. Such networks consist of computer systems designed to imitate, in streamlined form, certain basic principles of operation in the human brain. We report here on advances in one area of current research—optical signal processing using holography.

What kind of voice is this? Hesitant? Enthralled? Confident? Clearly, the last choice applies. The author projects a tone of self-assurance; he is in command of the material. He speaks as someone more knowledgeable, someone who wishes to teach, inform, guide. This is the persona of competence in science.

Of course, there are moments in almost every text where you need to qualify what is being said, to back off a little. This is frequently the case, for example, when you are making new generalizations, discussing the implications of your work, proposing new ideas, or simply admitting the limitations of what you've achieved:

> These results
> suggest the possibility that previous interpretations are erroneous . . .
> can be interpreted to indicate . . .
> support the concept of . . .
> are preliminary and indicate a need for further work in the areas of . . .

Knowing when to qualify what you are saying is part of being a confident author, projecting your expertise.

However, one of the most frequent complaints I have heard from editors is that beginning scientific authors are often far too hesitant in their writing. They are much more likely to say

> Our work here, though preliminary, may be considered to support the conclusion that . . .

rather than

> Our work supports the conclusion that . . .

The second example provides whatever "error bars" you may need. Scientists recognize such clues when they see them. Note that what the first example really says is, "Who are we to pretend that we can contribute something meaningful?" Or, as once put by a well-regarded investigator of the human condition (Groucho Marx), "Why would I ever want to belong to a club that would have someone like me as a member?"

Once again, look through your models. See how they set up their voice of authority. Pay close attention to places where facts or ideas are stated firmly and where they are softened by qualifiers. This, too, will help tune your ear and guide your hand.

ORGANIZATION

The skeleton of every document lies in its organization, the ordering of its parts and substance. Good organization involves several levels of order. First, there is the sequence of major sections, most commonly some version of abstract, introduction, background (e.g. geographic setting, materials and methods, previous studies), main discussion, conclusion, references. Second, there is the order of any subsections under these major headings. Together, the heads and subheads of a text should provide a kind of table of contents—indeed, it is sometimes a good exercise to extract them and see how well things are put together. Third, the writer must work out the degree of detail to be included in each section, as well as the progression of this detail. All of which sounds like a tall order, and it is.

Many writing guides provide one or another system for helping you organize your document. The basic idea is that structure determines flow of the argument, and therefore its persuasive quality. If this were true, however, we would be able to simply dump information on the page to tell our story. Organization can no more guarantee good composition than bones can bring a body to life. If we are to be physicians of the word, we require an equal or greater amount of biology.

No single group of methods, however carefully explained, can possibly encompass the range of needs, problems, and styles that scientific writers find individually relevant. There is no "right way" to impose order on the chaotic heap of information you've generated. Some scientists I know write best using a sketchy outline. Others require a veritable train schedule, laying out the arrival of one point after another. Still others plot one section at a time, as they come to it. Some writers, however, use no outline at all. One researcher of my acquaintance (a very skilled writer) works by pinning note cards randomly to the wall and then visually placing them into a dendritic (rootlike) structure. Different approaches succeed for different authors.

For inexperienced writers, there are a few practical techniques that might help in the early stages of composition. Making a provisional outline that includes a title for each major heading and subheading is one way to start, as long as you leave open the possibility that things may well (and often should) change as you write. If your research is laboratory-based, you might consider at least beginning with the standard structure—introduction, materials and methods, results, discussion, conclusions—and add subheads where you think they might go. You might choose only your major headings, place these on separate pieces of paper, and then write down ideas, data types, or the principal points you feel should be covered under each. If you prefer to plan things out visually, use a process similar to my friend the note-card user: write main ideas and data areas on separate cards and see what type of logical order they fall into. Another possibility is to assemble the illustrations (i.e. in rough form) you might consider using and search for a logical sequence among them that might guide the writing.

These are just a few suggestions. You should feel free to play with them in any manner you choose, ignore them altogether, come up with your own. In nearly all cases, it will take time and some experience for you to determine a style of organizing material that works well for you. It helps to be patient and to investigate a bit in this realm, too.

As an overall guide, think of your document as moving from the general to the more specific and then back to the general again—a kind of hourglass or vaselike shape. Begin broadly, that is, introducing your reader to the "problem" and its background, then discuss your methods, present your detailed findings, and finally venture outward again in your conclusions. The greatest concentration of details should be in the article's center, the discussion of what you found, what new data you generated or interpreted, what it means. In truly eloquent science, this type of hourglass approach tends to be repeated within each individual section of a document: the reader is repeatedly given a focus that works like a series of musical variations: general (allegro), specific (largo), and general again (finale: adagio). Functional communicators do not have to go this far. However, using the general-specific-general pattern to help organize your text is very effective.

Consult your models. Look at articles on similar topics, preferably in journals to which you might submit your work. If you need to, adopt or adapt a structure that seems appropriate or that might work as a beginning.

Please be warned, however: there is usually no way that you can decide all levels of order in your document beforehand. Most often, it helps to begin with a general plan, whether written down or not, and then proceed to write. In a few cases, your outline may hold true for significant stretches; more often, however, it will need to be altered. Writing, as I've said, is a process of experimentation and discovery.

STYLE

Style in science generally refers, not to the literary qualities of writing, but to the conventions governing its form. In other words, it refers to what is acceptable regarding basic structure, punctuation, capitalization, abbreviations, citation, reference format, and the like. A fair number of style guides are on the market in book form (the Council of Biology Editors' *Scientific Style and Format* is one example). Most of these are specific to particular fields and are helpful to editors. Many offer necessary advice about using certain universal aspects of style, such as metric notation, conversion factors, various constants, and the like. For everything else, however—that is, the great majority of stylistic aspects—nearly all these guides are of only partial use to the scientist-author.

I make this (heretical) statement for a simple reason: experience. Scientific disciplines are too diverse in their literature to obey any single manual of style. Standards and conventions vary at every level, among journals within a single field, indeed, even for single periodicals over time (e.g. between different chief editors). The reasons for this are complex. They have much to do with the interplay between institutional demands and the personalities and training of the people involved.

Editors may well go to war (and the grave) over the proper punctuation in a list or the form of references. But for writers, the practical result is obvious: because no final standards exist, you must take each journal on its own terms, which means examining closely the articles it publishes and, above all, consulting the "instructions to authors" or "suggestions to contributors" requirements for preparing manuscripts. This is the *only* way you can be sure to comply with these basic demands. It points up again the value of using models in science.

ENVIRONMENT

Your work of authorship can be lightened through your choice of setting. By this, I don't mean retiring to a country manor or mountain cabin, however attractive that may be. Instead, I refer here to self-knowledge: be aware of what environmental factors encourage and discourage you from getting to work. "Avoid everyday mediocrity in your working conditions," says Walter Benjamin, one of the great essayists of the 20th century. Some people need a clear desk, nice music (quality sonic wallpaper); others work better if surrounded by notebooks, reports, papers (the voices of data). Some prefer to write in their office, where everything is nearby; others work best at home, with a bit of distance and the motivating nag of family. Learn to know thyself; use your inclinations.

These are not trivial concerns. Writers have a number of tools to put in play; time and place are among them. As a human process, writing has an ecology to it, a personal dimension of engagement that can be nurtured or withered by physical context. The where and when of composition have importance, not just the how and what (we leave the why to the gods of employment and destiny).

GETTING STARTED

If you have trouble facing that blank page or screen in the beginning, here are some techniques to consider.

Try your title first. You probably have some idea of what you're writing about (this would seem necessary). Jot down a possible phrase, even two, or three (or more), to describe your topic, just to start.

Remember what was said about writer's block, near the end of chapter 3? Try some of the ideas listed there to start the flow of language in your mind. Read through an article or two on a similar topic. Be assured: this is *not* wasting time or postponing "the inevitable," but is valuable preparation and may well save time. The more fluent your inner voice, the more easily words will come to you.

Begin the paper with the introduction. If you can't think of an opening sentence, go to your models, choose an article on a similar topic, and imitate—even (to begin with) copy—the first sentence, substituting your own topic and terms. Note how the rest of the paragraph and the remainder of the introduction are structured. Emulate these too, if you find yourself still stuck. Then go back and revise, inserting background information specific to your subject.

Look at a previous paper of your own. How did you begin? See what type of generalization you used. Was it about the importance of the topic? Did you point out a gap in the existing knowledge? Perhaps you simply described the process that you studied or to which you added a new piece of understanding. Briefly analyze what you did and see if it might be useful for your present subject.

Sit down and discuss the subject with a colleague. This will force you to express it and may well plant some phrases in your mind that you can use afterward. Few things help clarify a topic in the early stages more than having to explain it to others. Take note of any questions your associate may have: as a listener, she is your first "reader" and can therefore aid you directly in determining what points to cover.

REVISING 1: FOR ORGANIZATION

Rewriting is where effective documents are made or lost. It is where good writers and their poorer relatives part company. There is a relatively simple reason for this. Revision is where you, the writer, get the chance to become a reader with the power of change. You see what type of experience the document creates. You see where it is rough and splintery, where it produces discomfort, where it is unfinished. At best, you read through it as if someone else had written it and given it to you to polish up (for a nice fee). You ask, at each step—as every author should—"Is this material that, if I saw it in print, I would want to have my name associated with it?" I'm not saying each sentence needs to be perfect; you're not writing *Madame Bovary* (Flaubert spent a month or more on single sentences and nearly went mercifully insane). But you want to avoid clumsiness, to sound articulate.

So far as I know, there has never been a Mozart of scientific composition. No one ever gets it right the first time—or, for that matter, the second. Revision is the chance we give ourselves to finish a document, to nurture it into maturity. Don't think of it therefore as "making repairs" or "replacing bad with good": rewriting is not how we "fix" but how we *complete* a text. Let's consider an example:

> Reefs of Silurian age have been the main source of oil and gas production in the state of Michigan for the past three decades. In 1990, the reefs produced about 25 million barrels of oil (84% of the state's production) and almost 132 billion cubic feet of gas (92% of the state's production). At the end of 1990, cumulative production from the reefs reached 211 million barrels of oil, 1.21 trillion cubic feet of gas, and 50 million barrels of water. Estimates of the primary recoverable reserves in these reefs are 300–400 million bbls of oil and 3–5 trillion cubic feet of gas. The main purpose of this study is to present a detailed analysis of the depositional history of one of the largest and best sampled Niagaran pinnacle reefs in the Michigan Basin.
>
> Pinnacle reefs of the Michigan Basin are isolated carbonate buildups completely encased by salt, anhydrite, and fine-grained carbonate deposits. Until the 1960s, the gravimetric method was the principal successful tool used for identifying pinnacle reefs in the subsurface. In more recent years, the search for reefs has been based almost entirely on seismic methods, with exceptionally good results. (Gill 1985, 123)

We can sense that this is not yet complete, both in organization and style. What should be done? It usually helps to look at organization first.

The introductory sentence seems perfectly fine, but it is immediately swamped by supporting numerical information. After wading through this, we jump suddenly to "the main purpose," which has to do with geologic history, not oil and gas productivity. At the start of the second paragraph, we shift gears again, to a statement about reef character, and, after that, skid into a completely different topic: methods for locating the reefs.

There is a lot of information here, but not all of it appears relevant, and most of it comes at us in disconnected pieces. What is necessary? What can we delete or replace? To answer this, look at "the main purpose" again. If we are interested, above all, in the geologic history of these reefs, do we really need to know how much oil and gas they produced in a particular year, or the historical methods used for finding them? Clearly no. So this tightens things up a bit. Do we want to get rid of *all* the numerical information given in the first paragraph? Probably not, because we need to add support and specificity to the opening claim, which establishes the importance of these reefs. Is this first sentence fine the way it stands? Look closely at the rest of the paragraph:

we don't learn these are *pinnacle* reefs, specifically, or that we are in the Michigan *Basin* (a particular geologic province) until the last sentence, where these very important terms are merely thrown in. Suppose, then, we revise the first sentence to look like this:

> Pinnacle reefs of Silurian age in the Michigan Basin are a main source of oil and gas, with estimated recoverable reserves of 300–400 million barrels of oil and 3–5 trillion cubic feet of gas.

This keeps the sense of importance that the original opening had, moves all defining terms to the front, and adds numerical support, all in a single sentence. Using our "main purpose" as an organizing principle, we delete the remaining numbers (we can put these back in a later section) and skip to a description of the basic geologic character of the reefs—this, after all, helps orient the reader as to what we are talking about.

> Pinnacle reefs of Silurian age in the Michigan Basin are a main source of oil and gas, with estimated recoverable reserves of 300–400 million barrels of oil and 3–5 trillion cubic feet of gas. These reefs consist of isolated carbonate buildups completely encased by salt, anhydrite, and fine-grained carbonate deposits.

Can we now add the last sentence in the first paragraph, to complete our introduction? Let's see if this works:

> Pinnacle reefs of Silurian age in the Michigan Basin are a main source of oil and gas, with estimated recoverable reserves of 300–400 million barrels of oil and 3–5 trillion cubic feet of gas. These reefs consist of isolated carbonate buildups completely encased by salt, anhydrite, and fine-grained carbonate deposits. The main purpose of this study is to present a detailed analysis of the depositional history of one of the largest and best sampled Niagaran pinnacle reefs in the Michigan Basin.

Not bad. But there is still a bit of a jump between the second and third sentences. The paragraph feels too short. And we have sort of tossed out the adjective "Niagaran," hoping that readers will do the work of definition themselves. What needs to be done? How about adding a sentence (shown in bold) to provide transition:

> Pinnacle reefs of Silurian age in the Michigan Basin are a main source of oil and gas, with estimated recoverable reserves of 300–400 million barrels of oil and 3–5 trillion cubic feet of gas. These reefs consist of isolated carbonate buildups completely encased by salt, anhydrite, and

> fine-grained carbonate deposits. **Previous investigation has established that reefs grew along the basinward margin of a major platform, which rimmed the Michigan Basin in Middle Silurian (Niagaran) time, and were later overlain by evaporite material (salt and anhydrite) due to a major drop in sea level (Hedberg, 1975; Corson et al., 1986).** The main purpose of this study is to present a detailed analysis of the depositional history of one of the largest and best sampled Niagaran pinnacle reefs in the basin.

We've taken some information from another part of the paper and put it here, up front. This is definitely better. We've added something that moves the focus from present-day character of the reefs to their geologic history, which is where we want to be to state our main purpose. We've defined "Niagaran," and we've also brought in some references. Notice, too, that our added information partly explains what was given in the previous sentence—in this case, salt and anhydrite are tied to sea-level drop. The transition is thus smooth and logical. Is there anything more we might do? How about making one more segue to the last sentence:

> Pinnacle reefs of Silurian age in the Michigan Basin are a main source of oil and gas, with estimated recoverable reserves of 300–400 million barrels of oil and 3–5 trillion cubic feet of gas. These reefs consist of isolated carbonate buildups completely encased by salt, anhydrite, and fine-grained carbonate deposits. Previous investigation has established that reefs grew along the basinward margin of a major platform, which rimmed the Michigan Basin in Middle Silurian (Niagaran) time, and were later overlain by evaporite material (salt and anhydrite) due to a major drop in sea level (Hedberg, 1975; Corson et al., 1986). **Such studies, though valuable, have been largely regional in nature.** The main purpose of this paper is to analyze in detail the depositional history of one of the largest and best sampled individual reefs in the basin, **the Belle River Mills feature.**

At this point, our introduction finally seems complete. We have inserted explanatory transitions that bring the reader from the importance of the subject to the importance of our work. We've moved from the general to the specific, and introduced needed terminology. We've revealed the gap in knowledge that our study is going to fill (always a good idea for introductions). We've identified our specific topic and described what we are going to do with it.

So we have a perfect paragraph, at last. Right? Hardly. No such fauna exists. What we have is a piece of writing that we can leave and move on from with a reasonable degree of confidence. If we had world enough and time, we could tinker with our intro a good deal more, ad infinitum. But then there would be the minor problem of never getting it published. Flaubert, of course, did work 20 years on *Madame Bovary*—but novels don't become dated quite so quickly as do data.

REVISING 2: STYLE

I've devoted much space to a single paragraph for three reasons. First, introductions are extremely important, especially to editors and reviewers—if you impress them here, you've done much to get them on your side (so that they may forgive other faults along the way). Second, I've tried to show how all the factors discussed earlier in this chapter come into play: trial and error, experimentation, thinking about your reader, organization. All these factors eventually need to work together. Third, and finally, our all-too-abbreviated example is meant to reveal how much decision making—identifying and solving of problems—actually goes into the rewriting process.

Now let's look at literary style in particular. This, too, requires close and concentrated reading. Note the following passage:

> As a method to generate low-density microcellular foam, we synthesized molecules that would dissolve in CO_2 under relatively moderate pressures, then associate in solution to form gels. Previous work has shown that gels can be created in traditional organic solvents through hydrogen bonding, association between ionic groups, or association between electron-donating and electron-accepting moieties. To form foams from such gels, it is necessary to preserve the supramolecular aggregates created in solution, both during and after solvent removal.

Right off the bat, we are faced with a dangling modifier ("As a method to . . . , we . . ."). Deleting the first three words provides a cure. Next, in the same sentence, confusion is created by that last comma and following phrase ("then associate in solution"). At first reading, this seems to relate back to "we synthesized"; on a second or third reading, however, we can see that it doesn't, and that it should form a parallel structure with "dissolve in," as follows:

> To generate low-density microcellular foam, we synthesized molecules that would dissolve in CO_2 under relatively moderate pressures and then associate in solution to form gels.

This reads better. Looking it over several times, however, it seems a bit redundant: if we say the molecules "would dissolve in CO_2," do we also need to say that they subsequently associate "in solution"? This is not a large point, but it allows for extra economy. Moreover, if we wanted to project confidence, as an author, we might also eliminate the conditional tense:

> To generate low-density microcellular foam, we synthesized molecules that dissolve in CO_2 under relatively moderate pressures and then associate to form gels.

That last phrase still hangs a bit. Reading the sentence through, we see that a temporal process is being stated: "that dissolve . . . then associate" What if we make this a bit more explicit:

> To generate low-density microcellular foam, we synthesized molecules that **first** dissolve in CO_2 under relatively moderate pressures and then associate to form gels.

Adding one word makes everything clear at last.

What about the second sentence? Here's a case where enumeration can help:

> Previous work has shown that a gel can be created in traditional organic solvents by one of three processes: (1) hydrogen bonding; (2) association between ionic groups; or (3) association between electron-donating and electron-accepting moieties.

Notice that I've inserted "one of three processes" instead of simply "three processes." As indicated by the conjunction "or," these processes are not simultaneous or overlapping. This is an example of adding precision and clarity for the reader.

As for the last sentence, a bit of reshaping can be done here too. First, the combination "form foams from such gels" sounds awkward. What if we changed it to read

> In order to produce a foam out of such gels, it is necessary to preserve the supramolecular aggregates created in solution, both during and after solvent removal.

Our final problem comes in the last phrase. What, exactly, does it refer to? Does it tell us that aggregates are created both during and after solvent removal? Or does it relate, instead, to the required preservation of these aggregates? As it happens, the rest of the paragraph eventually tells us that the latter is indeed the case, which we might have guessed. But the construction is unclear. Let's therefore change it:

> Producing a foam out of such gels requires that the supramolecular aggregates created in solution be preserved both during and after solvent removal.

This places the verb form "be preserved" closer to its true antecedent "solvent removal" and clears up any confusion. Our final version of the paragraph thus reads

> To generate low-density microcellular foam, we synthesized molecules that first dissolve in CO_2 under relatively moderate pressures and then associate to form gels. Previous work has shown that gels can be created in a traditional organic solvent by one of three processes: (1) hydrogen bonding; (2) association between ionic groups; or (3) association between electron-donating and electron-accepting moieties. Creating a foam out of such gels requires that the supramolecular aggregates created in solution be preserved both during and after solvent removal.

This is good, functional scientific expression. It is clear enough so that the reader will not stub his or her toe on any broken or confusing phrases. It is reasonably smooth, good in logic, and carries the argument in a particular, desired direction.

It will not have escaped your notice, I presume, that a fair amount of work was needed to get this far. In fact, I have streamlined the process, for the sake of brevity, leaving out a number of dead ends and unsatisfactory changes that I went through in my own revision of the above paragraph. It is important that you know this: perceiving what's wrong with style or organization does not mean that you can automatically correct it without a process of trial and error.

In the end, these are the sorts of questions you need to ask of your document and then answer through change. If you have any doubts about whether such change is needed in a particular passage, show your article to a colleague (or two)—this is a good idea in any case, particularly after you have finished a final first draft. There is no substitute for a foreign pair of eyes. Such foreignness (from the text) is partly what you yourself are after whenever you revise. If this is difficult to achieve, or if you become overly frustrated, put the article aside for a while (a few days, a week, if possible longer), and then return. You may find yourself the prodigal son.

REVISING 3: COMMON PROBLEM AREAS

There are, indeed, a very large number of possible stylistic and organizational pitfalls to which we are all prey. A tome of considerable mass would be required to delineate and address even half of them. Here I note only a few of the most common—those that seem to appear most often in published papers, books, reports, and proposals. Look these over and make sure you avoid being their victim.

The Run-on Noun Phrase

> Empirical ground-motion seismic attenuation data

or

> arctic tropospheric ozone depletion measurements

Stacking adjectives and nouns in this fashion reflects an attempt at increased shorthand, but it produces literary plaque. Any sentence or paragraph with this sort of structure is hard to get through. Though scientific language is dense, it cannot do away with readability altogether. Try, instead,

> empirical data on seismic attenuation from ground-motion studies

and

> measurements of ozone depletion in the arctic troposphere

If, however, the phrase is an accepted (standard) technical term, use an abbreviation, for example,

> high-mobility Si metal oxide semiconductor field-effect transistors (MOSFETs).

The Misplaced Definition

> Strong ground-motion information (estimates of peak amplitudes, durations, and phasing of seismic waves) is necessary to engineer earthquake-resistant buildings.

Definitions are best placed either at the end or in a separate sentence. Otherwise, they tend to jam or disperse the flow of reading. Try, instead,

> To engineer earthquake-resistant buildings requires strong ground-motion information, that is, estimates of peak amplitudes, durations, and phasing of seismic waves.

or

> Information on strong ground-motion is necessary to engineer earthquake-resistant buildings. Such information includes estimates of peak amplitudes, durations, and phasing of seismic waves.

Disjointed Topics

> Increasing global temperatures have been predicted to shift the geographic ranges of many species to higher latitudes or altitudes. Such increases in temperature have been tied to changing chemical composition in the troposphere, particularly the influx of carbon dioxide from combustion of organic fuels. Here we report on experimental evidence that slight fluctuations in water temperature control the impact of a key predator, the starfish *Pisaster ochraceus*. Preliminary analysis of *Pisaster* behavior indicates that it becomes inactive in low zone channels during periods of upwelling.

If the first sentence presents a good opening, the second jumps to a completely different topic. It should be deleted here and saved for a separate section, most likely the conclusions. The third sentence, meanwhile, relates to the first, but the last sentence presents another leap to a new topic and should be placed elsewhere, possibly in the following paragraph. Try, instead,

> Increasing global temperatures have been predicted to shift the geographic ranges of many species to higher latitudes or altitudes. Locally, communities may undergo changes in composition as species adapted to warmer temperatures are progressively introduced from lower latitudes. Here we report on experimental evidence that slight fluctuations in water temperature control the introduction of a key predator, the starfish *Pisaster ochraceus*.

In all cases, place similar topics or types of data together. Avoid "dumping" information on the page in the hope that the reader will make any necessary connections and thus do the hard work for you. Good organization involves grouping and ordering material. Go back and read some of your models with this in mind: see if they measure up.

The Land-Rush Opening

> Intense bursts of gamma rays that last up to several minutes have been observed during the past three decades and produce an optical afterglow believed to be synchrotron radiation from an expanding ultrarelativistic blast wave whose exact source geometry and emission mechanism is unknown but can be constrained by accurate optical polarization measurements.

Now take a breath. In good, functional scientific writing, the opening paragraph is divided into two basic parts: a short first sentence, which acts like an overture, and a series of following sentences, usually longer, that develop the topic into more specific variations. For the above, one might start with

> Intense bursts of gamma rays that last up to several minutes have been observed by many astronomers during the past three decades.

Or place it in the active voice:

> During the past three decades, many astronomers have observed intense bursts of gamma rays that last up to several minutes.

Then follow with

> When an optical afterglow is observed following such gamma-ray bursts (GRBs), they are interpreted to be synchrotron radiation from an expanding ultrarelativistic blast wave. The exact source geometry and emission mechanism for this blast wave remain unknown, but can be constrained by accurate measurements of optical polarization.

The opening paragraph should *introduce*—it needs to usher readers into the subject, not inundate them with a shower of topics and terminology. Move from the general to the specific, present key words in stepwise fashion, bring readers up to speed.

REVISING 4: PROCESS

What of practical methods or techniques in revising? The advent of computers and word processing has changed the process of rewriting forever. It is now possible to revise as we go along, leaving no tracks behind us. Technology has helped eliminate any obvious boundaries between writing and rewriting.

A very helpful procedure, particularly if you find it difficult to get started each time you sit down, is to read through what you've already done, get the material flowing in your head, then extend it forward. As you do this, you will probably find things to change before you get to your stopping point. Most people do revise every time they write. Something, somewhere, begs to be changed. This might well result in the earlier portions of your document being more polished than the later ones: try to be aware of this and to note if any important discrepancies in style exist.

To revise properly, you must be able to see your own work from a distance. This is where your models can come in again. Before going through your document to make changes, choose an article from your collection of good writing, preferably on a topic related to your own, and read it to get a sense of sound and style. You may need to do this more than once. Then look through your own work. How does it compare? Is it fairly smooth, logical? Where are there bumps or potholes? Where is it well done? What can you do to make parts of it read more like your model or your own higher-quality sections? Using your models of good expression as a measuring stick can be very useful in pointing up instances where you need to revise and, also, those where you don't. If the first will keep you working, the second will help give you the confidence to do so. Don't be shy about admitting your successes, even if they involve a single sentence here and there. Good writing, like a fine melody, is always an admirable achievement, no matter how locally it occurs.

CHAPTER FIVE

Writing Very Well

Opportunities for Creativity and Elegance

BEYOND THE FUNCTIONAL: "CREATIVE WRITING" IN SCIENCE

One of the most unfortunate folktales about science is that it has no room in its daily expression for creativity and personal eloquence. Except, perhaps, for a lecture by one or another towering genius (Richard Feynman is a recent icon) or a popular account by a researcher who, to the amazement of all, can "write excellent prose for the layman," scientists are presumed to inhabit a world utterly lacking in literary atmosphere, a planet not merely flat and functional, but bleak, distant, and cold. Indeed, the worst aspect to this bit of bad myth is that scientists tend to believe it themselves.

The truth, however, lies elsewhere. All forms of persuasive writing can be shaped creatively, with dashes of refinement, even beauty. Does this actually happen in science? Yes, indeed. Expressive distinction has not entirely escaped the gravity of the "scientific," by any means. It may be relatively rare—one article in a hundred, a few paragraphs in a lengthy review paper, scattered sentences in otherwise ordinary articles. But it is there, a figure in the carpet. Some observers have noted the fact: the famed paper on DNA structure by Watson and Crick (analyzed in chapter 2) has been called "a prose poem" by more than a few scientists. And though this kind of praise may well exaggerate the case, it still shows that elegance does not go unappreciated.

What does it mean to write elegantly, creatively, in science? Basically, it means producing texts that are not merely effective but *interesting* to read, those that offer a higher order of reading experience. In practical terms, this means fulfilling all the requirements of good, functional communication in a manner that is rhetorically sophisticated and inventive. There are many techniques that graceful authors employ, either consciously or intuitively. They include, for example, modulating sentence length and rhythm; posing questions at particular points in the narrative; applying parallel structure in individual sentences; employing refined turns of phrase; coining new terminology; providing smooth transitions between paragraphs and sections; and using syntax as an echo to meaning.

Such techniques represent a high level of expressive skill no matter what field or subject is involved. But in science, special circumstances need to be considered. As writers, scientists work within a system of expressive restraint and so must be relatively subtle in their creative acts. Conventions dictate that they not show off, that they avoid drawing too much attention to their language as an aesthetic medium. In the average technical paper or report, there are particular opportunities for applying elegance and invention. Let's look at some of them.

OPPORTUNITIES

Occasions for elegance occur in many parts of a document. Of course, any article or report can be made more graceful by excellent organization and polished writing throughout. But there are also particular points in an argument where good writing can be raised to a higher level with specific purpose. Here are some of them:

- Places where there is a need for generalization. This is a chance for adding memorable or inventive phrasing. Examples include the introductory and concluding sections of a document, especially in the very first and last paragraphs.
- Similarly, the first and last portions of each individual section. This provides an opportunity for generalizing gracefully and for providing subtle transitions, for example, those that pick up on points previously made or that forecast or suggest what comes next.
- Points in the argument that call for an added degree of emphasis. This may involve introducing an important, even unexpected, result; revealing a gap in the existing literature; stating a weakness in current theory; pointing out the (magnificent, unparalleled . . .) contribution of your work; suggesting areas for future investigation.

- Places of transition, where the argument takes a new turn, for example where it embarks on a new topic or direction, a different area of data or type of measurement.
- Areas in a paper or report where the language is descriptive and less technical, for instance where some aspect of history (e.g. of the existing literature or a particular technique) is discussed.

This list is meant to be illustrative, not inclusive. Different authors have often created different chances to assert their originality. What follows, then, is a series of examples aimed at giving instances applicable to all of the above, as well as some others.

EXAMPLE 1: INTRODUCTION AND CONCLUSION

Note this beginning to a paper entitled "Labyrinthine Pattern Formation in Magnetic Fluids":

> Several distinct physical systems form strikingly similar labyrinthine structures. These include thin magnetic films, amphiphilic "Langmuir" monolayers, and type I superconductors in magnetic fields. Similarities between the energetics of these systems suggest a common mechanism for pattern formation. (Dickstein et al. 1999, 1012)

This is a highly technical opening, yet with certain raw ingredients for eloquence. The first sentence is short and dramatic, the second follows by introducing specific topics, and the third reveals the fundamental logic underlying the investigation. Three brief sentences do a great deal of necessary work, in admirable order. Let's work with them a bit:

> Different magnetic fluids can form strikingly similar labyrinthine structures. In cases where similarity of structure reflects similarity in fluid energetics, a common mechanism may exist for pattern formation. Three examples for which this assumption holds true include thin magnetic films, amphiphilic "Langmuir" monolayers, and type I superconductors in magnetic fields.

Take a minute to analyze what has been done to the passage. Note the changes in sound (e.g. deletion of too many *s* words in the first sentence), the added parallel structure ("similarity of . . . similarity in"), and the new (previously hidden) logic in the second sentence. Note, too, sentence length: short, medium, long, which acts to draw the reader in and, at the same time, allows the introduction of increasing detail. Here, meanwhile, is the conclusion to the same paper:

> In conclusion, the branched patterns seen in experiment can be understood within perhaps the simplest dynamical models incorporating the competition between surface and dipolar energies. Taken together, the experimental and theoretical results indicate that an enormously complex energy landscape in the space of shapes can arise from a competition between short-range forces and long-range dipolar interactions in systems subject to a geometric constraint. A static theory of labyrinths would find only the minimum energy configurations, whereas the dynamic theories reflect the complexity of the landscape in the complexity of the labyrinths. (1015)

Again, we see the impulse toward eloquence, the desire for, and partial achievement of, making memorable statements. Let us, therefore, help things along:

> To conclude, we propose that the branched patterns seen in experiment be understood within quite simple dynamical models. Such models, in contrast to static theories, will be able to incorporate the observed competition between surface and dipolar energies. Taken together, both the experimental and the theoretical results indicate that, in systems made subject to a geometric constraint, an enormously complex energy landscape may arise from this contest between short-range surface forces and long-range dipolar interactions. Static theories cannot possibly account for such competition, but derive only the minimum energy configurations required for labyrinths to form. Dynamic theories are needed to reveal how complexity of the energy landscape is reflected in the complexity of labyrinth structure.

Once more, look at what was done. Very little by way of content was added or deleted. Instead, certain elements were reordered, sentences were divided up to give each individual point its own emphasis. A word here and there (e.g. "quite," "contest," "possibly") is employed for effect, without taking anything away from the content. Note, too, the passage is a bit longer than the original and, to purists, may even seem a bit redundant. An elegant conclusion, however, produces an effect that sticks with the reader: if perused, the above will reveal no repetition, but instead a degree of overlap that helps drive the central argument home. The paragraph ends, meanwhile, with what can only be described as a transparently veiled philosophical statement, a neatly packaged "grand principle" that all scientists would recognize as one of their own. Is there any reason to hold back from this sort of finale? Absolutely not—as long as you do it in the specific terms of your subject.

EXAMPLE 2: REORGANIZATION AND POLISH

Recall the following paragraph, part of which we examined in chapter 3 as an example of high-quality writing:

> Subduction of the Juan de Fuca and Gorda plates has presented earth scientists with a dilemma. Despite compelling evidence of active plate convergence, subduction on the Cascadia zone has often been viewed as a relatively benign tectonic process. There is no deep oceanic trench off the coast; there is no extensive Benioff-Wadati seismicity zone; and most puzzling of all, there have not been any historic low-angle thrust earthquakes between the continental and subducted plates. The two simplest interpretations of these observations are: (1) the Cascadia subduction zone is completely decoupled and subduction is occurring aseismically, or (2) the Cascadia subduction zone is uniformly locked and storing elastic energy to be released in future great earthquakes. Full resolution of this issue may prove elusive. Although it is somewhat surprising that no shallow subduction earthquakes have been documented in this region, the duration of written history is relatively short. (Heaton and Hartzell 1986, 675)[4]

As it happens, the paper does not begin with this elegant paragraph. Instead, it starts: "This is the third in a series of four papers that lead to an estimation of the seismic hazard associated with the subduction . . . beneath North America." The passage above comes third, after a discussion of the existing literature and the key aims of the article. To revise with polish, we would reorganize the introduction to begin with the above, follow with discussion of the literature, and end with mention of the paper's goals.

Can the above paragraph, meantime, be improved? Let's try.

> Subduction of the Juan de Fuca and Gorda plates has presented earth scientists with a dilemma. Despite compelling evidence of active plate convergence, subduction on the Cascadia zone appears to be a relatively benign tectonic process. There is no deep oceanic trench off the coast; there is no extensive Benioff-Wadati seismicity zone; and most puzzling of all,

4. In the succeeding examples, I have adapted parts of the article by Heaton and Hartzell to my purposes, revising the actual sections to varying degrees. Despite this treatment, it should be clear that this is generally a good example of an eloquent paper in the earth sciences.

> there have not been any historic low-angle thrust earthquakes between the continental and oceanic plates. The two simplest interpretations of these observations are: (1) the Cascadia zone is completely decoupled and subduction is occurring aseismically, or (2) the Cascadia zone is uniformly locked and storing elastic energy to be released in future great earthquakes. Although it is somewhat surprising that no shallow subduction earthquakes have been documented in this region, the duration of written history is relatively short. Full resolution of this issue may prove elusive.

In this case, I have done very little by way of revision. The word "subduction" or "subducted" was replaced in a few instances to reduce over-repetition of terms (very common in scientific writing). In the second sentence, the phrase "has often been viewed as" was changed to "appears to be," since, from the foregoing sentence, it is the process of subduction, not views of it, that forms the stated dilemma. The final two sentences have been reversed: this is not only good for logic, but also for sound and rhythm. Graceful opening and closing paragraphs usually begin and end with short sentences. On the other hand, nothing was done to the eloquent, even oratorical third sentence, with its wonderful parallel structure. Note how effectively and concisely the three points are delivered (three is the magic number here, stemming from an ancient rhetorical formula, as in Lincoln's famous "of the people, by the people, for the people").

EXAMPLE 3: USE OF QUESTIONS

Use of questions is a simple literary technique that is only sometimes employed in science, but it can be a graceful and effective accent that serves to add emphasis, introduce new material, or refocus a discussion. Posing a question adds a bit of drama:

> Cholesterol and related sterols are not uniformly distributed within the membranes of eukaryotic cells. Why is this so? Here we seek an answer by considering the effects of these flat, disklike molecules on lipid bilayers.

A question can orient and hold the reader's attention, break up an extended discussion, or turn the narrative in a particular direction:

> It has been observed that annual cycles of measles epidemics occur in areas where the birthrate is high. What are the limits to such a correlation and how can they be established? Previous models have assumed an average, therefore constant, birthrate, ignoring any effects that might result from seasonal variations.

There are also more restrained forms of asking questions, if the overt query seems too bold for your particular publication venue:

> Measurement of ultrashort pulses is a demanding task. The question has been asked: how can our existing instrumentation be improved? Our results suggest that . . .

Or, in an even more subtle style:

> Recent climatic modeling has revealed that maximum flux of anthropogenic carbon into the Northern Ocean occurs farther north than previous inventories would suggest. To investigate why this might be the case, we examined two mechanisms of transport . . .

> The onset of AIDS appears associated with an extension of HIV infection beyond the lymphoid organs into tissues such as the brain. Just how this occurs is unclear, but it seems probable that infected cells are among the responsible agents.

Note that the posing of questions does not have to be restricted to the opening portions of a document. It can also be appropriate to the main body of your discussion,

> During the analysis of experimental results, the question arose: what mechanisms might account for the single-file transfer of particles in this setting?

your conclusions,

> Is slip along the Cascadia subduction zone a benign process, occurring slowly as aseismic creep? Or alternatively, is elastic strain energy accumulating along this zone, and if it is, what is the nature of earthquakes that may result?

or even in headings,

> Cascadia Subduction Zone: Locked or Unlocked?

> Single-File Particle Transfer: Where and How Does It Occur?

Inserting questions into your document should be done sparingly. Queries of this type are forceful signposts for the reader: use them too often and you risk making your audience feel they are being manipulated (a great failure for any author). Place your questions in carefully chosen locations. As always, if you're experimenting with this technique and are somewhat unsure of its results, have a colleague read what you've written. Better yet, have two or three read it.

EXAMPLE 4: TRANSITIONS

One of the distinguishing features of any written document lies in the elegance of its transitions—how it knits together its various pieces, establishing a smooth flow between them. A simple technique for doing this is to announce, at the end of your introduction, what topics you will be discussing and then to proceed to do this in the same order mentioned, with each topic included in the headings of the following sections.

For example, the paper mentioned above on the Cascadia subduction zone and its potential for earthquake activity ends its introductory opening something like this:

> In this report, we extend the work of previous authors by systematically comparing trench bathymetry, gravity, and shallow seismicity for a world-wide sampling of subduction zones. It is hoped that by pursuing this line of investigation, one or more analogs may emerge for consideration.

The sections of the report that follow would ideally include topics like this: Trench Bathymetry and Gravity; Seismicity; Specific Comparisons; Analogs: Do They Exist?; Earthquake Potential along the Cascadia Subduction Zone; Conclusions.

Such linkage is needed in functional writing too, of course. Elegance enters in the logic of the general plan, but also in transitions that are given *between* individual sections. For instance, the first section (Trench Bathymetry and Gravity) could begin with the sentences

> The Cascadia subduction zone is unusual in that it has virtually no bathymetric trench. To begin to assess just how anomalous this is, we have constructed profiles of bathymetry and free-air gravity for many circum-Pacific convergent boundaries.

and end with

> We note, in particular, striking similarities in both bathymetry and gravity profiles between Cascadia and the subduction zones of Colombia and southern Chile. This raises the question of seismic activity.

These sentences obviously pick up on the opening, while providing flow into the next section (Seismicity). But note too that they provide a first answer to the opening query—how anomalous is Cascadia—and thus establish a question-answer pattern that might be pursued in each succeeding part of the paper. Indeed, the next section could then begin like this:

> A second apparent anomaly associated with the Cascadia subduction zone is the remarkable paucity of shallow earthquakes. While this seems puzzling for a major plate boundary that is undergoing 3–4 cm/yr of convergence, it bears asking whether and to what degree analogous conditions may exist elsewhere.

Note how several levels of transition are employed. First, we have carried forward the idea of (and the word) "anomaly" presented in the opening of the earlier section. Second, we now propose a possible contrast term, "analog," implicit in the final sentences of that section and mentioned in the last part of the introduction. Third, we emulate—without copying, but instead by adding variation to—the question-answer pattern mentioned above. Notice that Colombia and Chile, as possible analogs, are not mentioned. On the one hand, the relevant similarities must be established in relation to worldwide comparisons (this is simply good science). On the other hand, keeping them out here leaves the reader with a slight edge of anticipation, a heightened interest: will these same areas rise again out of the mix? Proceed, gentle reader . . .

EXAMPLE 5: WORD CHOICE

Choosing specific words is one area where inventiveness can be employed in an immediate way. The elegant writer consciously varies his vocabulary, trims repetition. At times, he will seek an unusual or unexpected word to create added interest. Look at one of the previous examples and a rewrite of it (bold type indicates changed words and phrases):

> In conclusion, the branched patterns seen in experiment can be understood within perhaps the simplest dynamical models **incorporating** the competition between surface and dipolar energies. Taken together, the experimental and theoretical results indicate that an **enormously complex energy landscape** in the space of shapes can arise from a **competition** between short-range forces and long-range dipolar interactions in systems subject to a geometric constraint.

> To conclude, the branched patterns seen in experiment can be understood within simple dynamical models that **embrace** competition between surface and dipolar energies. Taken together, both the experimental and theoretical results indicate that, in systems made subject to a geometric constraint, an **energy landscape of enormous complexity** may arise from this

> **contest** between short-range surface forces and long-range dipolar interactions.

Only a few words have been replaced, yet they give the passage a new elegance. The word "embrace" is fully acceptable and certainly no more ambiguous than "incorporate," but much more interesting and suggestive. Meanwhile, "contest" helps us avoid repeating "competition" and adds something extra, a note of struggle.

Here's another example:

> A central tenet of biomineralization is that the nucleation, growth, morphology, and aggregation of the inorganic crystals are regulated by organized assemblies of organic macromolecules. Control over the crystallochemical properties of the biomineral is achieved by specific processes involving molecular recognition at inorganic-organic interfaces. (Mann et al. 1993, 1286)

This is fairly clear, functional writing. We can appreciate here the use of the word "tenet" instead of the more common "concept" or "principle." But the rest of the sentence needs a little help, mainly because of word choice and word organization. Try this:

> A central tenet of biomineralization holds that the nucleation and growth of inorganic crystals, as well as their morphology and aggregation, are governed by structured assemblies of organic macromolecules. For a particular biomineral, crystallochemical control depends upon processes that involve molecular recognition along inorganic-organic interfaces.

Compare the differences closely. Nothing is lost in the change, except perhaps the attempted parallel structure between "inorganic crystals" and "organic macromolecules" in the first sentence, which didn't work in any case. We have replaced a number of ordinary words of general meaning ("is," "regulated," "achieved") with more graceful equivalents ("holds," "governed," "depends upon"), and have shortened a five-word phrase down to a two-word alliterative ("crystallochemical control").

Words are much more than linguistic bricks, to be piled on one another. Very often in scientific writing, we feel the need to repeat technical terms, almost to the degree of absurdity, as if they were such masonry. This feeling comes from the sense that there are no alternatives:

> Fracture analysis endeavors to measure the spacing and aperture of individual fractures, frequency of fracture occurrence, and the total extent of the fracture network.

This type of writing is, at best, barely functional. It is equivalent to a form of terminological dumping. Try, instead,

> Fracture analysis is an attempt to measure the spacing, aperture, and frequency of fracture occurrence, as well as the dimensions of the relevant structural network.

Adding elegance through word choice can also involve deleting certain unnecessary elements, cleaning up, in other words.

EXAMPLE 6: PHRASING

Refined phrasing in science? But there are many opportunities and much evidence for this. In fact, with its articulated demands for brevity and concision, scientific writing is one of the very *best* places for the felicitous turn of phrase. Note that the classical aphorism or maxim—a much admired literary model through the ages—derived much of its elegance from placing complex thoughts into highly confined form ("As the scale bends to a weight, so must a balanced man yield to circumstance" [Cicero]). While our aim in science is not necessary to set the literature aglitter with gems, we should recognize that creative phrasing is a common option, at times even with a drop of humor.

There are no final techniques for this type of writing that can be relied upon—except, of course, for emulating the work of others. Here are a few examples that I have collected in recent months:

1. A dearth of direct evidence urged us to search for the missing parameters.
2. Gravity waves in the lower atmosphere, forced by flow over mountains, have been observed and modeled for many years.
3. The role of the transfer matrix *t* is merely to modify this balance quantitatively as long as the lattice is perfectly periodic.
4. Blood vessels are the life-giving conduits that connect our tissues and our organs.
5. Research on thin polymer films has proven to be a drama of refound opportunity.
6. Multiple schemes for single measurements have too often yielded perplexing results.
7. During the years of its early development, the technique of synchronous laser pumping was advanced by a scattering of research teams.
8. The free energy of any physical system is rarely if ever free, but must instead be liberated by one or more conditions.

Notice in these phrases the use of sound and rhythm (examples 1–3), the artistic compression and neatness of word choice (4–6), and the employment of humor (7–8).

Such writing is excellently used as an occasional spice in a document, giving it flavor and finesse, raising it above the level of the ordinary. It is easy to overdo, however. Being too clever, or clever too often, will trivialize your subject. The best writers know to leave the reader affected and wanting more. To satiate completely, as Seneca says, is to erase an impression. A few, well-placed pearls will be sufficient for any single text.

EXAMPLE 7: METAPHOR

Contrary to what is often said, and far too often advised, the use of metaphor is alive and well in scientific composition. Scientific literature virtually teems with metaphoric terminology: white dwarf star, killer T-cell, RNA editing, plate tectonics, quantum charm, reporter genes, structural relaxation, molecular target, and so forth. True, these terms only begin as metaphors, when they are first coined. Over time, with standardized usage, they lose this charge and become identified with the phenomena they represent (i.e. they are no longer figures of speech in the active sense). But what they reveal, without question, is how strong and accepted the metaphorical impulse remains in science.

In what ways can this be applied to writing, in the ordinary sense? In fact, such application happens all the time, though usually in subtle fashion. Note this example once more (in its original form):

> In conclusion, the branched patterns seen in experiment can be understood within perhaps the simplest dynamical models incorporating the **competition** between surface and dipolar energies. Taken together, the experimental and theoretical results indicate that an enormously complex **energy landscape** . . .

Here are some others:

> Synapses are **focal points** of communication between nerve cells.

> One of the most dramatic events in the fossil record is the **explosive** diversification of marine invertebrates early in the Cambrian period.

> It was soon realized that determining the exposure history of individual grains was complicated by meteoroid impact **"gardening"** on the lunar surface.

> Convection in this case consists of nearly two-dimensional turbulence, with **meandering plumes** . . . that **drift** westward.

After reading these examples, look through your models and see where similar uses of figurative language appear. You'll note that it doesn't happen very often, only at particular points in an argument or document. On the other hand, for any given text, it may not happen at all—if the needed conditions do not arise.

Metaphors tend to enter scientific speech when authors reach for a descriptive word outside the normal pale of technical terminology. In some cases, this is done to fill a gap; in other cases, its purpose (consciously or otherwise) seems more imaginative and aimed at making the text more interesting. Writers might freely adopt a term from elsewhere in their discipline, from another field, or from ordinary speech. All of these approaches are entirely acceptable, provided that the chosen word or phrase is appropriate to the case.

In the above examples, figurative language appears in the form of both nouns and modifiers. One case (the third sentence) attempts (it seems) to coin a new term—note how the quotations serve to qualify the attempt, but also to let the reader know that the writer is being consciously inventive. New terms, of course, are being proposed all the time in science; this is both a necessity, as new phenomena are revealed, and an impulse, as individuals try to make a mark on their field. If you are lucky enough to be in a position to suggest a new term for your field, please do so thoughtfully, with a degree of intelligence (as a negative example, using comic-strip characters to name planetary features has the effect of appearing juvenile and trivializing). Such instances are revealing: they show us places where the wider culture enters into science in a direct way, therefore that scientific speech is hardly as flat and instrumental as is so often maintained.

IN THE END: LITERARY FINESSE IS KNOWING WHEN TO STOP

Just a brief final note. In nearly all cases, true authorial expertise in science lies in subtlety and restraint, not showmanship. There is certainly room here and there for play, whether this involves chiseling a suggestive phrase or coining a clever term. But on the whole, these opportunities are relatively rare. Science is reserved in its discourse; this is a historical condition, as I've said (chapter 2), but also a kind of knightly code.

Several times in the foregoing sections I've stated the importance of using one or another form of elegant writing on a selective basis. Let me here emphasize this once more. If you find yourself drawn to some of the techniques mentioned above and wish to experiment with them in a document you're working on, by all means do so with multiple blessings, but try to be conscious of the effects you're creating and how often you're creating them. Remember that anything you produce will eventually have to pass through the editorial gate, which can be quite narrow. If you have any doubts, show it to two or more colleagues and note their response (you may have to weigh this somewhat against what you know of their own literary inclinations). Editors and reviewers usually begin to object at the point where a document becomes too overtly personal, when "science" becomes subservient to self. The challenge for the sophisticated writer is to make his or her mark on a text, while drawing only momentary or background attention to the fact. This, in itself, is a considerable achievement. Elegance and restraint share bread, even in the lab.

CHAPTER SIX

The Review Process

Contents and Discontents

EDITORS AND WHAT THEY DO

Editors have one of the most difficult, thankless, and important jobs in all of science. The journal editor, in particular, is usually a scientist (one of us) and a volunteer, who gets no remuneration for his or her efforts, but who must keep up his or her own research, teaching, corporate responsibilities, and publishing attempts without skipping a beat, while somehow taking on the added jobs of literary manager, quality-control expert, and policeman of the field. This is the situation on the professional level. On the human level, meanwhile, the editor's work involves trying to minimize the indiscretions of others and, therefore, casting the stones of failure and success, hurt and happiness, on every side, all in the name of better science. Which helps explain why the editorial profession in science, like certain gases, is both noble and volatile.

Another point should be admitted. The strength and direction of a scientific field largely rise and fall with the quality of its editors. Bluntly put, editors have power and influence. Their specific work is to act not only as gatekeepers but as architects, to determine how wide or narrow a range of subjects will be accepted, which incoming articles are worthy of review, who will review them, what type of comments are to be made, whether these comments are to be accepted on face value, and what the final decision will be regarding acceptance, acceptance with major revision, or rejection. These are all critical steps in determining what science is published. Well-written articles on topics of significant interest are a mark of success, both for the authors and the editors. But the opposite is at least as true.

In most cases, an editor is guided in his or her work by certain protocols. For example, journals commonly have a bank of associate editors responsible for individual subfields, and to these people the head editor will delegate the responsibility for handling many of the papers that come in. Associate editors then choose specific reviewers or, in some cases, act as reviewers themselves. These days, reviewers are given standard forms to fill out regarding manuscript quality, and this helps make the final decision to accept or reject more straightforward. But the head editor nonetheless has the final say—the end responsibility always rests with him or her. The process is not democratic; at best, it is dilute plutocracy—imperfect, elitist, often inefficient, but effective.

Having the power to decide which papers are publishable means that an editor shapes both the scope and direction of a journal. Good editors are thus invaluable to a field and deserve far more recognition and reward than they get. The flip side to this, however, is that weak or overly autocratic editorial "regimes" can cause significant damage: even in the recent history of science, "quality control" has sometimes been an unfortunate euphemism for intellectual despotism. To curb such opportunities, and to prevent burnout as well, most editors are elected or appointed for brief periods, normally three to five years. In practice, a goodly number are *re*elected and *re*appointed, because few people wish to take on the burdens (and calumny) involved. This, too, is a reason that good editors are worth their weight in platinum.

THE REVIEW PROCESS: A STEP-BY-STEP OUTLINE

In practical terms, the review process in science serves to direct the revising of acceptable, written research. This means that, to editors and reviewers, all manuscripts are first drafts. As an author, you *must* be prepared to receive comment, criticism, and requests for revision.

When your manuscript is received, it is first logged in by an editorial assistant and possibly checked for completeness. If anything significant is missing—for example, you've included only one printed copy instead of the required two or three—it may be returned to you without even reaching the editor's desk. If the editor has to do this, you will have earned a black mark. Otherwise, your paper will proceed to take its place in the queue of submissions that await initial judgment.

The editor will then briefly survey the contents of the paper (title, abstract, headings, and illustrations, usually) and decide if it fits with the journal's scope. If it does, he or she next chooses who should review the paper or which associate editor in the relevant specialty should make this determination. Potential reviewers are then contacted, either by (e-)mail or phone, and the manuscript is sent to them. Along with the paper, a reviewer will receive a letter stating a preferred deadline for returning the manuscript, a guide for making comments, and an evaluation sheet, often titled something like "Confidential Reviewer Report."

This report has a list of questions that must be answered (yes/no, or good/fair/poor). Typical questions include, Does the paper present original scientific content? Has the material appeared in any previous publication? Are there significant errors in fact, logic, or argument? Is the title accurate and sufficient? Are the illustrations appropriate? Is the reference list complete? Can any portion of the paper be shortened or omitted without loss of content? At the end of the report are four choices, one of which is to be checked: publish as submitted, publish with minor revision, publish with major revision, do not publish. The first of these is there largely to keep the number of choices even; it is almost *never* used. More likely, if your work is at all significant and relevant and you have followed the instructions to authors, one of the two middle choices will be yours. If not, you still have options.

At this stage, the reviewers return their material to the editor, who must then make a final decision. Either the editor, or a managing editor or publications director, will then send a formal reply to the lead author. This will include a letter noting the decision with the principal reasons for it, a set of recommended changes, a requested deadline for resubmitting the paper (if it has been tentatively accepted), and the original printout copies of the paper with reviewers' comments on them.

The ball is once again in your court. You must decide where things go from here. To do this will require that you sit down and go over in detail what the editor and reviewers have to say. If your paper has earned a "publish with minor revision," then by all means proceed. If, on the other hand, you receive a publish with major revision, you have several choices. You can make most or all suggested changes and resubmit. You can accommodate some changes, provide point-by-point reasons (in your cover letter) why you find other recommended alterations unnecessary or invalid, and resubmit. Or you can officially withdraw your paper and look for another journal. If your paper is rejected, you can submit it to another journal as is, or you can use the comments provided by the editor and reviewers to reshape your material and then send it elsewhere.

Once your revised paper has been accepted, it will be copyedited to accord with stylistic conventions of the journal. The text will be set, made into page proof, and returned to you for final proofing. At this stage, you should make only the most necessary changes—meaning actual corrections—not updates of your data, new insertions, alterations to the structure of the article, or rewriting of entire paragraphs. This also goes for the illustrations when they arrive in final form (either with the text or separately). Despite the advent of digital publishing, changes at this point remain expensive. Too many will hold up publication of your paper or report.

POINTS AND POINTERS

Editors and reviewers are your first primary audience. If you can satisfy them, then the chances are good to guaranteed that your work will be reasonably suitable for, and will find, its larger, secondary audience—your (jealous) peers.

For this reason, whatever you do to make editors' task easier will likely prejudice them in your favor and thus increase (but never secure) the chances of acceptance for your manuscript. Editors very much appreciate certain signs of consideration. These include receiving articles that (1) are potentially within the scope of the journal or publication, and (2) comply with the instructions to authors regarding style and format. Neither of these two considerations is optional, for any author. If you fail to meet them, your manuscript is doomed, and, worse, you'll go down in memory as someone who wasted the editor's time (as well as your own).

Fortunately, these necessary conditions go together. Even before sitting down to write, or at least during the earliest stages, you need to decide where you're going to send your manuscript. If possible, you'll want to have a first choice and a backup choice as well. Selecting a particular journal (or other specific outlet), meanwhile, is the *only* way you'll be able to comply with the second consideration above, since different journals nearly always have different specifications. This is a brutal fact of scientific life. It would be much better—for everyone involved, but especially writers—if this were not the case, if standards were imposed across the board for individual fields. Perhaps some day they will be (though divine acts are rare). For now, however, take the time to find out the specifications for your particular journal or other publication and follow them. Note that such guides are usually provided for journals on the inside of every issue, and for other publications (e.g.

symposia transactions, annual volumes, conference proceedings) by the editor or editorial staff directly. It helps to copy these, or print them out, and keep them handy to refer to as you write.

I would advise, very strongly, that you not go to war, or even enter the field of conflict, over stylistic details. Editors and reviewers, being indelibly flawed humans like many of us, are prone to certain pet peeves regarding specific points of usage. Where one will always change "since" to "because" or "while" to "whereas," another will delete every occurrence of "the fact that" or "in consideration of." Copy editors, too, may be under orders to make alterations along these lines. Such changes are nearly always trivial—yes, unimportant. They derive from the great "schoolmarm" (Mencken) tradition in American letters and are based on bits of folklore about "proper usage." Except where they *truly* alter the meaning or make your prose flat and monotonous, you should simply accept (i.e. ignore) them and move on to more important things. The halls and stairwells of editorial offices across the land may well be stained red from battles over such positions, but it is blood needlessly spilled. Literary tics have been disobeyed by nearly every great writer (for wonderful examples, see the last chapter in Joseph Williams's book *Style*). You are better off without them and without worries over them. They will distract your attention and dissipate valuable energy better devoted to other things.

Regarding the actual submission of manuscripts, meanwhile, it's a good idea to keep up-to-date as much as possible with the electronic/digital forms that journals and other publication outlets encourage. It is clear at this point that such forms will, before long, become preferred and even exclusionary. This is because they are much faster, significantly cheaper, and generally more reliable and easy to manage than the current paper-based system. It is evident, too, that software will one day, not very long from now, take over from the pencil and pen in terms of changes and comments made on manuscripts (some journals have already begun to do things this way). Therefore, you would do yourself much good by staying current with these advances. Most current word-processing packages already offer such capabilities; try them out. The world of scientific publication is very definitely in a state of transformation.

One ugly fact for scientists: you can submit to only one journal at a time. Unlike in much of the humanities, simultaneous submissions remain unacceptable, *streng verboten,* in science. From the writer's point of view, this is a profound disadvantage—imagine spending several months or more writing a paper on a time-sensitive subject, several months more having it reviewed, provisionally accepted with major revisions, then, after a year has passed, finally rejected. Your material is now dated and must be reshaped for another journal, which might have been very interested in the first place, but must now also decline your paper, because a member of the competition has, during the interim, submitted an article on a similar topic . . . This type of situation does occur, from time to time, and can definitely leave scars. Fortunately, it remains the exception.

On the other side of the divide, the journal system in science would probably collapse overnight if it had to engage in overt competition for individual papers. To a degree, this is because of the need to safeguard the proprietary nature of research itself. The lack of simultaneous submission, that is, protects authors from having their work known and discussed by too many competing researchers. Obviously, this is no small consideration. The reality is that, as authors, scientists are both strengthened and weakened by the present system. All of this may well change, once the Internet becomes a prominent, perhaps the main, medium for publishing new science. Indeed, it is already changing for specific fields, as I will discuss in chapter 14. But much heavy machinery and bureaucracy needs to be moved. The journal remains king, both in the print and online worlds, and will probably continue its reign for some time to come.

THE DIGNITY OF A REPLY: NECESSARY ATTITUDES ABOUT CRITICISM

There is no doubt about it: dealing with the review process can be difficult, both emotionally and practically, but especially emotionally. Going over, point by point, what others have found inadequate in your written work may well be draining. It is especially hard for new or unseasoned authors, who have not yet developed an appropriately thick skin and whose work is perhaps more likely to earn a "publish with major revision" or "do not publish" reply. But it is likely to be difficult for almost any author, at some level. Scientists are not known for the graces of courtesy and tact when commenting on the work of others. You may feel judged, embarrassed, even humiliated—then angry, wronged, victimized. On top of this is the added time and effort that will be required to modify your paper or report, another reality that can inspire frustration.

These are very natural responses, and we are all prey to them. They are part of the rites of passage for every scientist, and they are sometimes hard to bear. Yet there are ways to deal with them that can help minimize their negative impact and maximize the benefits and maturity you can derive from criticism in general.

First of all, consider this truth: reviewers and editors have not offered their comments about you, the person, but instead about an inanimate, inorganic object—the manuscript. This object is an entirely separate reality: it, not you, went through the mail and sat upon a series of desks for evaluation. The manuscript came from your hands, certainly, but now it has a separate existence, and it is *this* existence to which all criticism is directed (a reason why the best criticism is always given in the third person). To the degree that you grasp this and keep yourself off the page, you will be able to evaluate the comments you receive with reasonable distance and balance. This is an extremely valuable skill to have as an author.

Second, understanding the separate reality of the manuscript will help make it clear that commentary by reviewers and editors is intended to make the paper better, more fully acceptable to a greater number of readers. Criticism is aimed *not* at a paper's destruction, but at making an improved piece of science. Granted, this is not how things are ordinarily expressed—usually, we must face the music of judgment: "The following aspects of your paper have been deemed unsatisfactory for the reasons given." Yet the goal is not to chastise, but instead to raise the level of what you have done to a more elevated stratum. Remember that the primary task of the editor is to ensure the quality of his or her journal, which is only as high as that of the articles contained within it. He or she therefore truly wants your paper to be as relevant, well-written, and complete as it can be (within certain constraints, of course), so that the light will shine that much more brightly on both of you.

Third, criticism is inevitable. Please accept that this is true, not just for beginners but for absolutely everyone. I myself (and others I know well) have written hundreds of papers, articles, and reports, received rewards for some of this work, and even been accused of eloquence, yet every single paper I submit to a peer-reviewed journal comes back loaded with comments, with demands for change in content, style, sometimes even organization. The great majority of these comments are intended to improve the paper, not to impugn or belittle its author. Reviewers take their job seriously. This means that, like editors, they look at any paper assigned to them as a kind of first draft, a creation that can always be made better (which it can). They offer their comments because, in part, they want to participate in your work, to help make it better. Yes, there are times when this can go too far. A reviewer may make so many suggested changes, for example to sentence style, that it seems he or she is trying to take possession of the paper and rewrite it entirely, in his or her own image. A good editor will be alert to this sort of thing and intervene in some way (if only in the form of a note to the author), so that the (real) author is not insulted and his or her time is not wasted. Reviewers who are repeatedly guilty of this type of oversuggestion should be warned by the editor and, if necessary, dropped. Like other adults, reviewers need to know how to control themselves.

These considerations should help make it clear that you do not have to accept every suggested change to your manuscript. On the contrary, if certain criticisms appear invalid to you, let the editor know, by all means. You should send a separate letter explaining your response, point by point, with your resubmission, after making other changes you feel are relevant. In the rare cases (and they are, indeed, rare) where a reviewer may have overstepped the bounds of professionalism and allowed personal reasons (politics, vendettas, hobbyhorses, grinding axes, etc.) to dictate criticism, the editor is obligated to send the paper to another evaluator, and he may well strike the original reviewer from his list in the future. Maintaining quality requires a degree of vigilance in all aspects of the review process.

If your paper is rejected, do not despair—act. Find another journal. This is easier said than done, of course, since you'll have to reshape the paper in certain ways. But it is an essential response, both for you and your field. Editors frequently return the manuscript with a brief discussion of why it was rejected and, occasionally, what might be done to make it acceptable. Use this as helpful advice for submitting elsewhere. Alternatively, consider using portions of what you've written as the kernel for a different paper, appropriate for another journal. Cannibalizing what we've already done is a fully legitimate, and often practiced, form of expanding our options.

As a successful scientist, you are likely to be a professional author—writing is an essential part of your professional responsibility—and therefore you need to conduct yourself professionally in all situations. In all your correspondence with editors (you should never be put in touch, directly, with your reviewers), and with anyone else involved in the review and publishing process, it is *absolutely essential* that you keep your cool, remain courteous, and speak to the point. This will not only serve you in good stead in all your external dealings, it will also help you achieve a certain useful distance from things. The authorial voice of professional calm is an extremely useful lie. No matter how heated a situation may become, it will impose dignity and help you stay in control. Using this voice is how you can best defend your work and uphold an image of command and competence. "Comments by this reviewer appear biased, unhinged, and irrelevant" is not the sort of response an editor will feel inclined to accept. Rather, couch what you say in tones that appeal to reason, that will make an editor feel he or she is being addressed as an intelligent, rational third party: "Comments by this reviewer, though well considered, are invalidated by the following points."

Being a professional means, above all else, communicating like one. Rage if you must, bestride private vales of smolder and fume, but stick to the high ground and cooler climes in your dealings with others. Moreover, on a wholly pragmatic level, keep a paper trail, that is, copies of everything that may pass between you and a publisher, so that if anything ever goes seriously wrong, you have the evidence on your side.

In the end, to complete the circle, perhaps the most practical advice of all rests in the arena of expectation. If you send in your paper knowing it will draw criticism and will have to be revised, you are definitely ahead of the game. This may not be easy. Secretly, at some level, we all hope to be that lucky (but apocryphal) exception whose article or report is immediately accepted, and accepted intact. The truth, however, as I've said, is that nearly all articles submitted are first drafts to the review process. Therefore, plan for revisions—emotionally and temporally.

Finally, consider this fact: acceptance rates for major journals generally range from 20% to 65%. Periodicals like *Science* and *Nature* represent the lower range of this spectrum, as might be expected. Much higher percentages (>50%), however, are typical of the major specialized journals for individual fields and, especially, subfields. For example, Paradis and Zimmerman (1997, 191) indicate that American Physical Society journals, of which there are more than a dozen, accept on average 9,000 manuscripts out of a total 15,000 submitted each year, thus no less than 60%. In my own field of geology, some of the most prestigious periodicals, such as the *American Association of Petroleum Geologists Bulletin* and *Geological Society of America Bulletin,* regularly accept between 55% and 75% of submissions. Informal surveys (not only my own), meanwhile, suggest that as many as 80–90% of all papers offered to publishers eventually find a home in print, in some form. Despair over rejection or required revision, however understandable in the short term, is not a practical long-term response.

CHAPTER SEVEN

The Scientific Paper

A Realistic View and Practical Advice

YESTERDAY AND TODAY

The modern scientific article, though the core of technical communication today, began life discreetly, as the lecture and the letter. These were transcribed in the earliest journals, the *Journal des Sçavans* in Paris and the *Transactions of the Royal Society of London,* both of which appeared in the 1660s as formal outlets for presentations given before each respective society, the Académie des Sciences and the Royal Society of London. What is striking about these early articles, to the modern eye, is their splendid variety. It is a wonder to see what teeming heterogeneity is there, a medley of subject and approach, from laboratory research to speculative thought experiments, from microscopic observations to field reports on distant lands. The science of 300 years ago could be expressed only through such diversity. It contained in vitro what would eventually grow into the most prolific enterprise of knowledge production the world has ever seen.

To such variety, the scientific paper has remained ever true. No single definition can encompass its full reality. "A written report describing original, replicable research" is fine as a high-altitude description, but nothing more. The two critical terms here—"original" and "research"—change considerably in definition across disciplines. Recall that there are as many as 50,000–70,000 refereed scientific journals in current publication (Paradis and Zimmerman 1997, 178). A bit of variety might therefore be expected. In my own field, geoscience, everything from mathematical simulations to descriptive fieldwork counts as research and has its place in the literature. Other disciplines would recognize neither of these species as acceptable or relevant, but instead would focus on laboratory effort. Science is no more a single method to reveal truth than is art.

Thus, it is *essential* for every scientist to explore the literature of his or her own field. This is really the only way you can gain a realistic idea of what counts as publishable research and what doesn't. It means learning what constitutes the "primary literature" and the "secondary literature," and what might occur along the increasingly porous boundary between. The world of scientific publication is both conservative and malleable and is now in significant flux. Investigating the literature is therefore an important part of your research.

TYPES OF PAPERS

Scientific publication is actually a vast, evolving cosmos today, one that matches the breadth and diversity of technical effort itself. Many different types of "papers" now exist in the world of scientific publishing, some results-oriented, some not. Such variety also reflects the fact that many journals (actually, editors) like to remain flexible in terms of what they can offer their readers. In practical terms, this means that there are a number of different kinds of articles wherein a scientist may discuss work in progress or express opinions about the work of others. The following list gives just a few of the more common article types.

Results-oriented papers frequently include

- *The Research Article.* This is the mainstay of technical publication, an actual report of new work intended to introduce new knowledge in a specific field. Depending on subject and focus, this article can vary from a few pages to 20 pages or more. Those of greater length may be referred to in a journal as "reports," "articles," "original papers," or some other term.
- *Letters/Short Communications.* These are brief research papers, usually less than five or six printed pages. They may carry material that is peripheral to the main thrust of a journal, but deemed of direct interest to readers. Other terms used for this type of article are "research notes," "brief communications," and "short papers."
- *Commentary/Forum.* This category is usually a brief, debate-oriented discussion. It may expand on earlier articles, seek to clarify or amend specific points, or offer criticism based on new work

by the author(s). Some journals routinely provide space for limited, formal debate and will include both a critique and a reply. Other common titles for this type of writing are "discussion/reply," "opinion," "viewpoint," and "debate."

Types of journal articles that are less results-oriented may include the following:

- *Letters to the Editor/Correspondence.* Most premier journals carry a section with this title, containing technical responses to earlier published material. Such letters usually serve the purpose of debate and criticism. Less often, but increasingly in some quarters, they respond to an editorial, book review, or other results-free writing.
- *Review Article.* This type of report presents a critical survey or overview of recent research and thinking in a particular field. It is aimed specifically at keeping scientists up-to-date with findings in their own and related fields. Such articles are commonly longer than a research report and contain a large reference list. However, it has become fairly common for journals to publish "minireviews" on topics of high current interest. Review articles tend to be commissioned by journal editors.
- *Book Review.* Many premier journals carry a book-review section, plus a list of "books received." Such reviews are usually quite short and follow a style and format specified by the journal editors.
- *Editorial.* A small but increasing number of journals publish opinion pieces, usually one page long, on topics that might range from recent discoveries or ethical concerns to federal budget cuts and public controversies. In the past, editorial writing was the exclusive province of editors; however, it is a growing trend to encourage scientists to contribute to this forum.

As I say, these are among the most prevalent types of articles in scientific journals. But they hardly define the entire field. There is significant variety well beyond the lists above. Moreover, journals are themselves highly diverse in what they offer: some regularly provide examples of every article type just mentioned, while others publish only research papers. Most lie somewhere in between.

In all cases, therefore, it is essential for you to survey the journals in your field to see what, exactly, the range of your options is regarding publication. All of the examples given above are put through a review process of some sort. They can all be found in bibliographies of research papers and should be listed on any curriculum vita. In short, they usually count as part of the "primary literature," which I take up in the next section.

PRIMARY, SECONDARY, AND OTHER LITERATURE

Traditionally speaking, the "primary literature" constitutes the first publication of original research results, commonly in a well-recognized journal. "Secondary" refers to any subsequent appearance in print of such results, in such venues as review articles, conference proceedings, book chapters, and so on. Primary publications are peer-reviewed and count directly toward tenure or research posts, both inside and outside of academe. The secondary literature, in the past, was usually unreviewed, ephemeral, and commanded far less prestige. It was viewed as much less important to career advancement.

Over the past several decades, these distinctions have begun to soften. This has happened in direct response to the changing, indeed broadening, realm of scientific publication as a whole. Primary results now appear not only in premier journals, but also in proceedings of important one-time symposia, special volumes on designated topics, monographs, government reports, book chapters, and more. New journals, too, are being launched continually. The position of these is usually uncertain for the first several years, but might be high from the beginning if they are the first to cover an emerging subfield.

The secondary literature in science, meanwhile, has come to include a huge array of publishing opportunities that, again, vary considerably from one field to another. Abstracts, transactions, conference proceedings, local bulletins, posters, newsletters, Web sites, and other such outlets normally count in this category. In recent years, these outlets have become more important, as the lag time between acceptance and appearance in primary publication has grown—a gap of one to three years now typically exists. For work in highly competitive, frontier areas, this is no small consideration. The secondary literature helps stake a claim for future primary publication by offering avenues for partial release of research results or general description of them.

By far the most significant challenge to the older ivory-tower journal system, however, with its cumbersome lag time and high-priced access, is Internet publication. This has rapidly and unevenly become a major medium for dissemination of research throughout the scientific world, and there is little doubt that it will continue to do so in the future. All this provides more options for the scientist-author. It is necessary, therefore, that you explore what this medium has to offer in your own discipline, where it stands, what outlets it offers, how you can use it to best advantage. Whatever new opportunities the Internet may provide, however, it is clear that the journal article, in something approximating its present form, will remain the nucleus of the *corpus scientia* for at least the near future. Because of this, we should take a look at how it is put together and what goes into it.

ARTICLE STRUCTURE: PARTS AND THEIR PURPOSES

It bears repeating that there is no standardized template for scientific papers applicable to all journals and all fields (nor should there be). To convince yourself of this, go to your nearest university library and browse through periodicals from different disciplines. You'll find that, while some require a fairly set article structure, many others do nothing of the sort.

Most journals, however, require that papers contain some version of the following parts: title, abstract, introduction, background, methods/ materials, discussion of results, summary/conclusion, references, acknowledgments. What follows is a brief review of each of these parts.

In reading through this, and in your perusal of the literature, remember that each major section in a paper centers on a distinct type of content and, to a significant degree, uses a different style of writing.

Title

The title is the most important phrase or sentence (except, of course, for all the others). Certainly, it is the most-read portion of any paper. It announces the article, telling other scientists if they need to read the paper or not. Try to keep it short; use key words that will help your paper be indexed properly (otherwise, it may be lost to its potential readership). Titles are usually phrases. In some publications (e.g. the premier journal *Cell*), sentences and questions are acceptable. Read through several issues of the publication to which you intend to submit for models to emulate. If you're having trouble choosing a title, try jotting down several working versions during the writing of the paper, then revise at the end. Here are some examples of good titles.

LABYRINTHINE PATTERN FORMATION IN MAGNETIC FLUIDS (short and attractive)

THE SOURCE AND FATE OF MASSIVE CARBON INPUT DURING THERMAL MAXIMA (dramatic and to the point)

A ROLE FOR PROTEIN PHOSPHATASES IN LONG-TERM DEPRESSION OF THE HIPPOCAMPUS (well-phrased)

LABORATORY MODEL FOR DEEP EARTH CONVECTION: HOW IMPORTANT IS A THERMALLY HETEROGENEOUS MANTLE? (states both the topic and the "problem")

Poor titles, meanwhile, might look like this:

TO CREATE A PROTEIN-BASED ELEMENT OF INHERITANCE (Who or what is doing the creating?)

VOLTAGE-DEPENDENT LIPID MOBILITY FACTORS IN THE OUTER HAIR CELL PLASMA MEMBRANE (Noun phrases too long; try VOLTAGE-DEPENDENT MOBILITY OF LIPIDS IN THE PLASMA MEMBRANE OF OUTER HAIR CELLS.)

AN UNNATURAL BIOPOLYMER (Too little information.)

A SUSTAINABLE ROUTE TO THE CREATION OF MICROCELLULAR MATERIALS USING CARBON DIOXIDE TO PRODUCE FREE-STANDING GELS WITH LOW BULK DENSITY AND NANOMETER-SIZE CELLS (Too much information; save everything after "materials" for the abstract.)

Abstract

The abstract is the second most read portion of any paper—and, increasingly throughout science, a crucial publication in its own right. Indeed, abstracts are doubtless *the* most widely exchanged and distributed type of formal scientific writing in the world today. They are often the only published evidence of conference talks, presentations, and research updates. They are frequently excerpted and republished in reference volumes. They are now included in most online bibliographic databases, a major new aid to research. And abstracts are also forms of knowledge "capital" that scientists trade among themselves almost as readily as they do greetings (or criticisms). All of which highlights the considerable importance of this unique, condensed form of written communication.

If, as the saying goes, brevity be the soul of wit, then the abstract requires clever chiseling. A good abstract is more than an executive summary or a series of generalizing statements. It is much closer to a minipaper, a compressed version of an article or talk (which, in a sense, justifies its separate publication), minus figures and tables. Think of the abstract, therefore, not as an add-on but instead as a stand-alone, an entity that, if decapitated from the rest of the paper, would convey its bodily substance. In many cases, after all, it will be all that a reader sees.

This may sound formidable. It needn't be so. Writing good abstracts doesn't depend on gifts from above, but on observation (of good examples) and practice here below. Try to follow the basic order of points in your article. In many cases, one to three topic sentences for each section of your paper are sufficient. Be sure to include scope and importance of topic, basic approach used, some specific data, and most important conclusion(s). Keep abbreviations to an absolute minimum. Don't include too much hard data (it clots the narrative)—select only the data that help establish the "problem" or support the main conclusion, perhaps just enough to tickle the readers' interest so that they might go on and read or search out the full text.

Beware of using word-processing software that "autosummarizes" your paper: this type of tool remains quite primitive and mechanical, and you may well find that the result requires more editing and rewriting than if you'd done it yourself in the first place. A few experiments with it should tell you whether or not it might be helpful. But don't mistake it for a shortcut. There are no technofixes for good writing.

Perhaps the most common problem in creating abstracts is the urge (doubtless felt by all) to cram everything in and get it over with—an impulse that, when allowed its day, will lead to long, heavily burdened sentences that need to be read several times, before true confusion sets in. For example,

> Analysis of historical and recently available 2-D reflection seismic data along the eastern and northwestern margins of the San Juan Basin reveals a close relationship of Laramide-age basement faulting with fracture orientation in the vicinity of several fractured Mancos Shale reservoirs, here used as a basis for an important recent structural study of oil occurrence in the basin.

It makes good sense to begin an abstract with a short sentence—a brief and clear statement, in fairly *general* terms, giving the importance of the subject or focus of the study:

> Seismic data reveal a close relationship between basement faulting and the orientation of fractures responsible for oil production in parts of the San Juan Basin.

Or

> We present results of a seismic data study confirming a close relationship between Laramide-age basement faulting in the San Juan Basin and oil production from fractured reservoirs.

Abstracts need to be just as readable as a well-written article. Here too, do unto your reader as you would have him or her do unto you. Think about what information is absolutely necessary, and what isn't:

> Biologists now believe that the first colonization of land by eukaryotes resulted from symbiosis between a photosynthesizing organism (phototroph) and a fungus.

A nice opening sentence. But "phototroph" doesn't appear anywhere else in the abstract. It isn't needed and, in fact, adds a distracting wrinkle, because it signals the reader to look for further mention of this term. Delete it and the sentence is excellent. Here's another example:

> For some time now, there has been considerable experimental and theoretical effort aimed at understanding the role of normal-state phases in high-temperature cuprate superconductors.

Again, a reasonably good opening. But the abstract is no place for "conversation." Try, instead,

> Considerable experimental and theoretical effort has been aimed at understanding the role . . .

In *all* cases—this is very important (which is why I say it so often)—check issues of your targeted journal(s) for examples of what is accepted. Study the instructions to authors, as they might have information that pertains specifically to abstracts (e.g. length and format).

Finally, be aware of an unwritten rule followed by some editors. This rule states that the abstract should not repeat verbatim any sentences in the main text. I seriously doubt the average editor has the time or inclination to check this in detail for every paper, but he or she may be tipped in the direction of doing so if, for example, the first sentence of your abstract is identical to that of your introduction, which follows immediately after. Moving a few phrases around should be sufficient to take care of this.

Introduction

Much more than a rhetorical welcome, the introduction is an essential entry hall into the house of your paper. It needs to offer the reader a hospitable and substantive reception. This means stating the problem or topic you are writing about, why it is important, how you have approached it (in general terms), and what is new about what you have done. A good introduction thus points up the gap in existing knowledge that your paper will help fill. In most cases, it also provides a degree of essential background, for example, an outline of existing knowledge on a topic, definitions of terms, or a brief review of the existing literature.

Good introductions begin with a brief sentence that launches the topic. This is immediately followed by supporting details:

> The development of chemotherapeutic agents for the treatment of HIV-1 infection has focused primarily on two viral enzymes: reverse transcriptase and protease. Regimens including agents directed at each of these biochemical targets are effective in reducing viral load and morbidity and therefore mortality. However, the long-lived nature of the infection and the genetic plasticity of the virus have made it apparent that new antiretroviral agents are required to deal with the appearance and spread of resistance. To address this issue, it may be important to consider the process by which viral DNA achieves insertion into the host cell genome, namely integration, which is catalyzed by HIV-1 integrase. (Hazuda et al. 2000, 646)[5]

> Subduction of the Juan de Fuca and Gorda plates has presented earth scientists with a dilemma. Despite compelling evidence of active plate convergence, subduction on the Cascadia zone has often been viewed as a relatively benign tectonic process. There is no deep oceanic trench off the coast; there is no extensive Benioff-Wadati seismicity zone; and most puzzling of all, there have not been any historic low-angle thrust earthquakes between the continental and subducted plates. (Heaton and Hartzell 1986, 676)

A common way to close an introduction is with a description of what you did, plus one or two major conclusions that resulted; for example,

> This paper suggests a new approach to identifying inhibitors capable of preventing HIV-1 integrase catalysis . . . In particular, diketo acid inhibitors were found to manifest significant anti-viral activity as a consequence of their effect on integration.

> In this report, we extend the study of previous authors by systematically comparing trench bathymetry and shallow seismicity for a worldwide sampling of subduction zones. Such comparisons lend support to

5. I have slightly altered the opening to this article. In the original, the second two sentences are combined into one: "Although regimens including agents . . . , the long-lived nature of the infection . . ."

the possibility of great subduction earthquakes off the Pacific Northwest coast.

Background

A background section is frequently needed to bring readers fully up to speed for the material that follows. Depending on the journal or publication venue you've chosen, this may or may not be a required part of your article. The background portion can take a number of different specific forms. In some cases, it is where a "review of the existing literature" or "previous investigations" should go. In other instances, for example when you performed fieldwork, a section that describes your setting might go here. Depending on the specific subject, many disciplines require a theoretical section to help outline major assumptions, previous thinking, current hypotheses, mathematical bases, and the like. The only template to follow is what the publication requires. If this is unspecified, decide whether you actually need a separate section or whether you can roll the relevant information into the introduction.

Methods / Materials

A methods section is required for studies that involved some type of laboratory work, whether based on experiment or measurement or both. A number of different titles are regularly given to this section, depending on the field: Materials and Methods, Experimental Methods, Research Procedure, or Apparatus and Procedure, among others. All of these titles, however, are used for a separate portion of a paper or report that offers a description of the tools and techniques you used to solve the "problem" stated in the introduction. As such, it defines an essential part of many papers and reports for several reasons. First, it allows the reader to follow and, if desirable, repeat what was done—thereby permitting verification or questioning of the stated results, a fundamental tenet for all scientific research. Second, it permits an evaluation of how skillfully the work was designed and carried out, thereby offering a possible model for other researchers. Third, it places the work in a certain historical context—stating that these were the tools and the techniques that were available at this point in the history of the field—and therefore leaves the work open to fertile re-study when newer methods come along (as they inevitably do).

For these reasons, the writing in this section must be especially straightforward, even cookbook, in style. This section requires a specific type of discourse not found in other portions of the article (and thus points up how complex a document the scientific paper really is). As always, consult your models of good writing. Evaluate how they handled their particular subject; gauge whether you might be able to repeat their experiments or not.

Discussion of Results

In the discussion of results, you say what you found and what it means. By this time in the article, readers will be prepared for your specific contribution. They will have been briefed on the topic, on what gap in knowledge it will fill, and on how previous work has handled or overlooked it. They will also have taken a short tour of your laboratory or field setting and have been shown how you chose to go about your investigation. Now it is time to return to the lectern, present your data, and reveal its significance.

For laboratory studies, and sometimes for other types of investigations as well, this part of the report is frequently divided into two sections, results and discussion. In this case, your data should be presented and briefly discussed in the first section, with your major interpretations reserved for the second.

For nonlaboratory work, these two sections are sometimes used, but it is also common (in some fields, far more common) for authors to craft individual headings specific to their topic. In this case, findings and interpretations are often merged under each section. For example, individual data areas (e.g. seismic information, gravity data, magnetic data, geologic structure) may have their own sections and subsections, each of which might include the presentation and interpretation of the relevant information.

If this is the approach in your field, examine closely how results similar to your own have been put down effectively by other authors. Pay attention to the use of illustrations and how these collect and summarize information. Most often, you will probably have some idea of the types of tables, charts, graphs, maps, and images you might want to use. But a glance through your models or the literature in general might reveal specific forms that you overlooked and that suit your topic very well, or, alternatively, that suggest how you might creatively adapt a given form to your case. If you are using a results-and-discussion structure, beware of the impulse to slosh information onto the page. Too often, inexperienced writers want to rush through the data section in order to get to the interpretations (the exciting part). But data are never just "data"—numbers, measurements, quantity. They have meaning, obvious or not. Any display of information will contain or suggest interpretations. These first-order interpretations are really part of your results.

Let us look at some examples:

> The three tests performed as part of the initial experiment revealed an improved correlation between the root mean square of factor Y and the observed trend Z.

> Six of the RNA functional groups were found to make significant energetic contributions to the formation of the signal recognition particle complex.

> It is clear from Figure 3 that stresses are rotated up to 45° in the lower crust, due to deep detachment faulting beneath the Basin and Range province.

Any of these would do well in a results-type section. Yet notice that they offer interpretations, too.

What then of the discussion, or its equivalent? Here, you would speak more generally:

> Results derived from the series of experiments reported here suggest [that previous correlations involving factor Y have underestimated the importance of . . .] or [the following conclusions: (1) . . . , (2) . . . , (3) . . .]

> Our studies on the structure of signal recognition particle in *E. coli* provide a new understanding of conserved ribonucleoprotein elements. [Follow with specific points.]

> As shown here, both analytical and finite-element models of extension in the Basin and Range support the conclusion that rheological differences between the upper and lower crust determine fault geometry at depth.

Once you've put down your major conclusions, take a look back. Make sure your tables and graphics are appropriate and that they aim in the direction you want them to, that is, toward your final interpretations. Keep in mind the experimental nature of the writing process. Be open to the possibility of discovering new interpretations after you have graphed and presented the data and had a chance to look at it anew, in these forms.

Finally, the discussion of your results is a good place to express confidence. Talk about the larger implications of your work, whether these relate to theory, methods, or analyses. Don't be afraid to make a claim for the importance of what you've done—after all, you've worked hard and deserve notice. But don't overdo it either. Einstein never claimed to be "Einstein" (read his papers and you'll see). Watson and Crick stated only that the subject of their DNA paper "had considerable biological interest." Therefore, gauge the import of your work relative to that of the subject and what others have said. The two "infinities," self-negation and grandiosity, are the Scylla and Charybdis of scientific writers.

Summary or Conclusions

In a summary or conclusions section, you have an opportunity for bringing closure. Every article, whatever its story, requires an ending. In science, this usually means several paragraphs that briefly summarize what was done, what was found, and what significance the work may have for present and future research. This final section should not repeat data or go over detailed interpretations given earlier. Instead, it should link your work to larger ideas and issues.

A good concluding section provides balance to the introduction and a sense of progress. In your final paragraphs, you should return to the research topic with which your paper began in the first several lines and state what new knowledge or new thinking has been added. In other words, the topic remains, just as important as before (perhaps more so), but illuminated or expanded or refocused in a particular way. Second, recall that your introduction began generally and progressed to more specific points—your conclusions section should do this in mirror fashion, ending with the broadest statements.

Here's an example showing both these points, from the first and last lines of a recent article:

> The composition of a planetary surface is an important indicator of its early evolution and subsequent chemical alteration. The surface composition of the jovian moon, Europa, for example, has been modified in a number of ways over time, including through intense bombardment by jovian magnetosphere particles . . . that could have effected change through radiolysis. The relative importance of this chemical alteration process has not yet been established for Europa . . .
>
> The abundance of H_2O_2, and the existence of an Na and O_2 atmosphere thought to be produced by energetic-particle bombardment of the surface, demonstrate that surface chemistry on Europa is dominated by radiolysis . . . Predictions, characterization, and identification of surface chemical species on this planet must therefore consider radiolysis effects more closely than in the past. (Carlson et al. 1999, 2062–2063)[6]

6. The quoted text has been slightly altered from the original.

An effective conclusion tells the reader that something has changed, that science has been increased in some way.

End Matter

End matter includes acknowledgments, references, and appendixes. The first two of these are required ingredients for any paper. They are the places where you most directly reveal the social reality of your work—where you show that you are part of a scholarly community and that you have depended upon the kindness of grant-giving strangers (or, in some cases, friends).

When writing the acknowledgments and reference list, consult directly your targeted journal for style guidelines. There are no final standards in this area: different journals typically employ different styles. Thus, checking the journal is the *only* way you can be sure you are doing things correctly. Acknowledgments offer gratitude to both granting institutions and to a possible range of individuals. If you're uncertain who to thank, look at examples from published papers to see who might be included.

The reference list is an essential element to get right. Most instructions to authors specifically note how the editors want you to write this part of the article. Follow their guidelines closely, to the last semicolon and period: this is a mark of being professional and considerate (much time, money, and emotion are wasted by copy editors in this area). If no stylistic guidelines are provided, refer to published papers in the targeted journal as models. It is a good idea to do this in any case to get an idea of how long your list should be and what types of sources are generally used. Some journals—not very many—allow for explanatory notes to be included in the reference section. Once again, consult other articles to see how this is done.

Appendixes are occasional additions to papers. They are usually reserved for detailed offerings of data (especially in tabular form), mathematical derivations, discussions of innovative methods, or other aspects important to the work being published, yet that would take up too much space (overwhelm or distract the reader) if included in the main portion of the article. An appendix is thus an opportunity, reserved for very special cases, to present more information than commonly occurs in the brief space allotted a journal paper. Some journals allow subsections within an appendix; others require that you divide these into separate appendixes or shorten everything as much as possible. If you have any questions about whether you should include an appendix or not, check with several experienced colleagues or the editor of the journal.

OUTLINES CAN HELP

Because the scientific article is usually firmly structured, outlines can help in the actual writing process. There are many approaches to building outlines—I've touched on some of them previously (see chapter 4, "Organization"). But one of the simplest and most direct methods is to begin by writing down, in provisional form, the major headings for your various sections, for example, introduction, methods, sections for each data area or subtopic, conclusions. Or, if you can't decide on these yet, even tentatively, use the main parts of the article: introduction, background, discussion of results, conclusions. List the topics or major points you might want to include under each of these, in any order you want, as ideas come to you. Then see if you can find a logical order for them, within each section. Sometimes—often, really—if you get this far, it will be enough to get you started writing. As always, if progress lags, consult your models, look through other papers in your targeted journal. It very often happens that papers and reports can be organized as hybrids based on other documents. Don't be shy to emulate or, selectively, even copy what others have successfully used: after all, they inevitably did the same. Above all, leave any outline open to change: it is a guide, not a template.

A cautionary note is in order. Many scientists (and nearly all manuals on scientific writing) make the error of assuming that results and discussion define the true core of any report. This is like saying that the torso is the most important part of the body. Without the other parts, it is meaningless and dead. All sections of a competent article are necessary and require the same amount of care and attention; there is no specific hierarchy among them. A careless introduction or inept conclusion are the writer's equivalent to bad lab procedure or misleading data. As scientists, we are taught that data and interpretations are our main business, certainly. But when we present our work to others, we become writers and speakers, and communication has its own demands.

CITATIONS IN A SCIENTIFIC PAPER: THEIR MEANING AND USE

Citing other authors in a paper does several things. First, it offers accountability. It tells the reader that you are familiar with the most recent, significant literature in your area and that this literature has aided you in your work. Second, citation is a way to outline a community of like investigators—a collegium, if you will. Third, citations are a tool by which you express various degrees of agreement and disagreement toward the work of others within this community: colleagues can be cited favorably ("the excellent work of Barnes et al. 1987"), unfavorably ("Delpy [1994] failed to consider"), flatly ("has been the subject of numerous studies, e.g. Batts 1978; Resin et al. 1983; Foresby 1985, 1992"), and in qualified fashion ("the work of Jensen et al. [1998] requires further support"). Most documents employ several of these types—they are how scientist-authors rank their cohorts and competitors and position themselves toward them. Fourth, citation is also a way for making certain claims to originality or, perhaps inadvertently, the very opposite.

These are complexities that flow from one another and that we all recognize in science, usually on an intuitive level. I make them explicit here to show, again, how multidimensional a scientific document really is, but also to help provide some awareness of the reasons for the way scientists write as they do. Here, too, that is, you can gain an added degree of control by understanding what is being done. Let's take several examples.

Case 1. Suppose, in the interest of being thorough and precise as a good scientist should, you are abundant with your citations, adding a reference to support every generalization, making sure you've included all relevant sources within the past five years, no matter how obscure. Perhaps you've just completed your PhD and are writing it into article form: you have all these worthwhile references that you've worked hard to compile and properly place—why not use them? Isn't this part of being complete, conscientious, and professional? The answer, I'm afraid, is no. Think of your references as a type of data: to be (or not to be) professional means being selective, offering your reader only as much as she or he needs to reconstruct your work without any sidetracks or distractions. Piling up citations makes you look hesitant, timid, and worse, derivative. On the practical side, it also renders your document varicose, clotting it up, making it physically difficult (thus inconsiderate) to read.

Case 2. Suppose, on the other hand, that you decide to include only a very few citations, to only the most well-known literature: what then? Another backfire. This type of referencing (or lack thereof) will suggest tones of intellectual plagiarism. The desire to appear foremost and original will make you seem guilty of fraud, a risk no good scientist should be willing to take. If your work truly is pathbreaking or exploratory, make this known in how you present your results and how you discuss the work of others. The height of unprofessionalism is to pretend that you have few or no peers.

Case 3. Let us propose, finally, the circumstance where an author cites his or her own work fairly often, or perhaps very often, as well as that of other colleagues: what blessings or sins arise here? In fact, self-citation is one of those unavoidable gray areas in scientific writing that can't be easily dictated either way. Referencing your own work too often (say, more than 20% of the time or so), or stacking your reference list with your own articles (e.g. more than 5 or 6 out of 30–40 total), will produce an image of unprofessionalism for the same reasons just given above. If you are one of the only researchers in a particular area, however, you may be forced to grant yourself a high degree of recognition. Even here, a touch of humility defines the better part of valor: at all times, try to strike a balance between your own contribution and your debt to others.

One final matter of practical importance. During the past several decades, the practice of "citation analysis" has become a significant factor in the minds and careers of some scientists. This type of analysis, which remains controversial, involves counting and ranking citations in the technical literature and then using the data to define such moral-political aspects of science as patterns of influence, emerging frontier areas, university or laboratory standings, international comparisons, and so on. The complexities of why and how scientists reference each other (only a few of which have been noted above) render this type of approach debatable, even dubious, in some of its conclusions.[7] Nonetheless, a number of institutions have taken to using the results in decisions regarding tenure and advancement. This is both unfortunate and, to a degree, predictable (as a kind of technofix shortcut). Certainly it encourages an erosion of ethical authorship. Whether or not it has led to an actual increase in the frequency of self-citation and "buddy citation" is difficult to say (it probably has). Practically speaking, as an individual scientist, you should be aware of how your institution views such "data."

COAUTHORS: HOW TO ORGANIZE AND MANAGE CONTRIBUTIONS FROM OTHERS

Research today is collaborative, and most scientific articles have more than one author. Indeed, it is now common for papers to have at least three coauthors, and some include as many as a dozen (the record, in fact, is over 100). This should make us stop for a moment and think

7. For a brief but astute discussion of this topic, see Fuller 1997.

about what "authorship" means in this context—for it is clearly different than in other areas of intellectual work (imagine a literary essay with seven authors . . .). In nonscientific fields, having your name on an article indicates that you did most of the research and, above all, the actual writing: the idea of "the author" is still tied directly to the crafting of language, the putting of words to paper. Not necessarily so in science. Authorship here is much more loosely and variably defined to include all those who have made an important contribution to the work being represented, whether this be in the design and execution of experiments, the overview and mentoring of work, or the actual writing. The masthead of any individual scientific paper is therefore closer to a dramatis personae or list of participants.

This brings with it a series of decisions that are not always easy to make. Friendship, collegiality, laboratory or departmental politics, institutional traditions: all these can play a role in determining who is to be included and in what order. Generally, the first name given is recognized today as the senior author, with progressively decreasing contributions assumed for the order of names thereafter. This is sensible, yet it may not be true. After the first several names, how can you decide who goes where if ten more "authors" are involved? Diplomacy and negotiation—or allegiance to alphabetical order—are often required. Authorship, like research, is not a democracy, but it shouldn't be a dictatorship either.

Realistically, there are no final rules in this area—but a few guidelines can help. Fairness is a good start (and finish): include as authors only those who have actively participated in the intellectual work being represented. This means those who generated real substance, who could accurately be said to have conceived and worked through a certain portion of the research and writing and who might therefore be able to defend it. It may help to define, precisely, what "authorship" means for *your* paper—what range of contributions, what degree of involvement in the research, what level of participation in the actual writing. Above all, to thy own collaborators be true: if a (prestigious) colleague contributes an idea or two at a seminar or during a hallway chat, does this merit his or her inclusion alongside those who have worked hard for months in the lab or field? Probably not, even if the idea proved pivotal in helping direct or focus the research. Instead, use your acknowledgments section to recognize this type of contribution.

At some point in your career, however, you may have to deal with authorial hitchhikers (dare we say parasites?). This is a lamentable, even deplorable, reality, but it is one that flows from the reward system in science, where publications count as career capital. It is still an occasional practice for the names of lab directors, department heads, graduate advisers, or other dignitaries to be added as a matter of course to any paper that emerges within their jurisdiction, whether they had any substantive input or not. Another example is where a certain lab may routinely include the names of technicians on all papers. If and when you encounter such a situation, it is necessary, for your own career, to carefully evaluate the situation. Every article is born out of a series of social situations (this is very much part of science, too), and it is advisable to be a savvy traveler and participant in these. Pick your battles carefully: live to write another day.

CHOOSING A JOURNAL

How do you decide where to send your work? *International Bulletin of Ornithology* or, instead, *Ozark Birdwatcher's Newsletter*? Prestige and wide distribution are frequent considerations. You'd like your research to make an impact, to reach as large an audience as possible. But it sometimes pays to be realistic, too.

Experienced scientists will likely have a journal or two in mind at the outset. If you are an early-career author, however, you may need to consider this question more thoroughly. Certainly you need to pose it at the very beginning of the writing process, since different journals have different stylistic specifications. In truth, assuming you're reasonably familiar with the literature in your field, a good part of the decision will be made for you: such aspects as the subject, approach, and scope of your work will tend to focus your choice down to a few publication possibilities. Beyond this, look for as close a match as you can between the content of your research and what has appeared in these publications. Which journal might be most welcoming of your research (properly written, of course)? Where have colleagues working in the same field, on similar projects, published their results? Where have articles appeared that you might consider using as models or guides for your own? What about any new journals—including electronic (online) journals—that might be appropriate to consider? Such questions might help focus or finalize your search.

The nature of scientific publishing requires, however, that you choose one, and only one, journal at a time. Again, each publication has its own blueprint for manuscript style and preparation, and science does not allow simultaneous submissions (i.e. sending your paper to several journals at once). If you've been rejected by one journal and are considering submitting to a second choice—after appropriate revisions, of course—make sure that you modify your paper to fit that journal's specifications. You can also look at a rejection as a new opportunity: think about breaking the original paper up into smaller pieces and publishing them separately, for example as notes or as letters that offer comment on articles that have appeared previously. This is a fully legitimate and well-recognized outlet for research. It can also help lay the ground for resubmitting your original paper (again, in modified form) elsewhere.

PRACTICAL CONCERNS: OPTIONS FOR SUBMITTING THE MANUSCRIPT

Beginning in the 1980s, science entered a new phase in its media of communication, and this has provided a set of new options for submitting material to journals and other publishers. As things now stand, you can choose to send papers and illustrations in hard-copy (paper) form, electronic form, or some mixture of the two.

Scientific authors should be particularly aware of the different options now available for electronic submissions, as these are rapidly becoming the norm in many fields. Relevant options include sending material on diskette, on large-capacity (e.g. 100 megabyte) disks, on compact disc (CD), or as e-mail attachments. Many journals allow you to send the text in electronic format and illustrations in hard copy, or vice versa. This is fine, at least for now. But it is advisable that you learn how to submit everything electronically, as this is clearly the direction that science (and publication generally) is headed. The most rapidly growing type of submission is through e-mail: as science continues to make expanded use of the Internet, it is probable that more and more journals will move to this medium, due to its cost savings and greater ease of use.

Electronic formats for submission bring with them new responsibilities. Authors must pay close attention to specifications regarding word-processing and graphics software, word count, document formatting, use of color, and so forth. Editors and referees look closely at the legibility and completeness of a manuscript, both of which can be compromised if file-format problems arise. Some journals have begun to provide reviewer and editorial commentary in the form of embedded annotations, which word-processing software now supports. It is likely that this too will expand in the future, particularly as online journals become more important.

CHAPTER EIGHT

Other Types of Writing

Review Articles, Book Reviews, Debate/Critique

ADVICE AND CONSENT

Research papers are the core of scientific publication, to be sure, but the entire fruit is much larger than this. Most scientists know as much from their weekly reading experience. Premier journals today, and certainly the big international periodicals like *Nature* and *Science,* often carry a range of writing: review articles, letters to the editor, editorials, book reviews, meeting summaries, debate, obituaries, news of the profession, and more, in addition to research reports. Increasingly, in fact, journals have broadened the array of writing within their borders, or given new importance to types (e.g. the review article) that formerly appeared only occasionally. Such variety exists, in part, because of increased pressure to keep periodicals attractive and useful. Call it a policy of noble survival. Editors have consciously decided to make their publications responsive to the evolving communicational realities and needs of science—in essence, to make the journal itself even more a marketplace of ideas, a nexus of professional expression, delivery, and exchange. Not all journals have made such a decision. Some still publish only one kind of article, perhaps two. Science has much room for different visions of what a periodical should be and do.

But the point is this: every type of published writing in science is a distinct contribution to the field. Of course, it is common, and entirely understandable, that you might focus on getting your own research published above all else. Yet nonresults writing provides another avenue, at times more relaxed, for playing a significant role in your discipline. It is another opportunity for adding valuable content and (let us not overlook) placing your name before peers and the public, including job review committees.

There is also the question of influence. The skilled scientific author, who publishes review articles, book reviews, and so forth, as well as research papers, is likely to have a larger professional presence. He or she will reach a greater number of readers and achieve higher levels of visibility and authority than the average scientist. The growing openness of many journals to different forms of writing has the effect itself of opening up new space for skilled writers to take center stage, to expand their contribution, increase their reputation, and gain status for their work. Print and online journals today offer more chances for writers to engage in productive self-interest. Indeed, even for those who would rather focus on research per se, some types of nonresearch writing—especially review articles and correspondence—can be used to further expose work already done or to pave the way for its fuller publication. No surprise, then, that those in charge of job decisions, whether in academia, industry, or the government, look upon such publications in a very positive light, and why scientists eagerly, and justly, list them among their published works.

What follows below are some basics regarding this "other writing," specifically its most common representatives: the review article, book review, and debate/critique (of another's work). Be assured that these types of writing, too, vary considerably between fields and journals. As with research papers, there are no universal standards—except the one advising you to explore what the journals of your choice have to offer. In every case, it will help greatly to identify and use models from the literature.

THE REVIEW ARTICLE: FUNCTION AND ROLE

An essential part of every scientist's work is keeping up with his or her field, and, to the degree possible, related fields as well. This is no simple task, especially these days. But it is exactly what the review article is intended to help you do—summarize and evaluate recently published work on a specific topic of importance. Such a topic is usually broader than that of a research paper, for example, "Evolution: The Role of Human Influence" (instead of a single example) or "Aromatic Metal Clusters: Bonding and Stability" (rather than analysis of a particular compound) or "How Do Thermophilic Proteins Deal with Heat?" (not a discussion of *Protentis caloris* occurrence in a Galápagos sea vent).

In many cases, a scientist is drawn to write a review article because, in the course of normal research, she or he has surveyed the literature in some detail and has some specific perceptions and opinions about it. Many scientists, in fact, carry around inside them review articles, like novels, waiting to emerge. The major difference, of course, is that reviews are easier to write and are likely to be more interesting and valuable to read (plus, you don't need an agent).

A good review article can do several things. It can (1) provide an idea of the current state of knowledge and, possibly, its historical development; (2) discuss recent and new directions in research; (3) point up gaps or limitations in this work; and (4) make suggestions for future research. Not all reviews, to be sure, do all these things—some focus only on the first or second tasks. But some type of perspective still must be added. As such, this type of article is really the closest thing to a scientific essay, a form of intellectual appraisal. Indeed, in some part, what literary or art criticism is to the humanities, the review article is to science.

Over the past few decades, reviews have become invaluable to many disciplines, particularly in the life sciences. Keeping abreast of new developments is no longer a simple or easy task. The reasons are several: growth in the number of journals, thus in the volume of published papers; an ever increasing variety of new disciplines; deepening specialization within individual fields; and an expanding trend toward multidisciplinary research. These realities have everything to do with how science itself is advancing in the present age. The effect, however, is to make it more challenging and necessary to stay current on new thinking and discovery. I say this not to suggest that the reader feel bound to a hopeless situation (on the contrary!), but instead to highlight the practical importance of the review article. While reviews are more common in the biological and medical disciplines than in the physical sciences, this seems likely to change, for all the reasons given above.

As with research papers, the review article comes in different shapes and sizes, depending upon the publisher. Some journals publish reviews up to 20 pages long, meant to survey a topic and discuss where it might, or should, be headed. Other periodicals offer "minireviews" of five pages or less, intended to hit the moving target of a fast-paced area. Still other journals provide both types. There are journals devoted entirely to reviews. But reviews are also collected in book form, as in the well-known *Annual Reviews of . . .* (medicine, ecology, geology and geophysics, etc.) and various yearbook series. Thus, as with any other type of writing you may be interested in doing, by all means explore your field.

Reviews Are Commissioned

Review articles are usually commissioned. This doesn't mean that only editors and their immediate family are ever able to contribute. It means that prospective authors need to submit an article proposal for consideration. An effective proposal usually contains a topic synopsis, an outline (annotated in most cases), and a description of the number and content of any figures. If the editor accepts your proposal, he or she is requesting the article—in professional terms, it is "commissioned" or "assigned." The process is intended not to erect barriers, but to protect both editor(s), who must deal with considerable literary mass in any case, *and* authors like yourself, who might otherwise spend months laboring away on an article with little chance of acceptance. Having your proposal accepted, however, doesn't guarantee publication. There is the small matter of submitting a well-written, well-organized manuscript, which then goes through the same basic process of peer review and required revision as a research paper.

Your first step, even before launching into your proposal, should be to contact the editor to see if he or she is interested in your topic. This is the time to ask for any specific guidelines in submitting your proposal (form, specific content). Be assured: editors *always* appreciate being consulted in this way at the beginning. It is a form of professional courtesy, and as such, good diplomacy. Editors also understand that a review article can do two things for your career. It can establish (or confirm) you as an expert voice among your peers—no small achievement, in any field. And it can make you the source of controversy—because, in most cases, you are interpreting and even judging, whether directly or indirectly, the state of knowledge in your discipline. Such considerations mean that reviews are often carefully edited. As always, the ultimate goal is to produce a better article and thereby serve the readership.

Important Points for Writing a Review Article

As an essay of sorts, the review can be more fluid in its structure and style than a research paper. But for this very reason, it needs logic and flow at least as much, if not more. Most reviews have only two standard parts in the main body, an introduction and a conclusions section. Between these two end points, you have a great deal of latitude about how to organize your article, how many sections to use, what you call them, what you put in them, or how long they might be.

There are several basic approaches that authors use to write a review. Here are the most common.

Alternative 1. In the *historical approach,* you choose to outline the development of knowledge on a chosen topic. This may involve charting the emergence of a particular subdiscipline or research area. To do this, you should discuss the milestone studies in chronological order and point up the more important findings, advances in method, and principles that have emerged from this work. Normally, this involves going back no more than several decades. However, if you are concerned with theoretical advances, with the maturation of ideas on a fairly broad topic, you may need to review concepts that go further back than this (e.g. on a topic in evolutionary biology, it may be useful to call upon one or another implication of Darwin's original theory).

Alternative 2. In the *experimental focus,* you concentrate on methods and materials, on recent trends in experimental work. This might involve answering such questions as, What aspects of the topic have been most avidly pursued? What methods have been used or developed to study the phenomenon, and what are their advantages and limitations? Have any problems been revealed in these methods? What forms of data analysis are used, and do they also have limitations (are there any important links between experimental methods and analyses of results)? What might be done to possibly improve research in this area?

Alternative 3. The *concepts and hypotheses* approach focuses on the state of knowledge, as intellectual content. What are the reigning hypotheses about the topic and who has framed them? How might these concepts relate to the larger theoretical framework in the field? To what degree is current knowledge explanatory, predictive, or descriptive? What is the situation of debate, as it presently stands? What positions have been taken, on the basis of what data? Have there been important, recent challenges to otherwise accepted concepts? Are certain reconsiderations under way? What significant questions remain unanswered, or not yet pursued?

Alternative 4. The *implications* approach seeks to discuss the latest developments and to outline their possible consequences. Questions here include, What are the primary advances with potential for practical application? What types of application might come from this new work? How might these advances improve existing technologies or apply them in new ways? What potential do they hold for new therapies, new and better predictive capabilities, or other applications? What important gaps or limitations need to be addressed in order that this potential be realized? How might this be done?

Alternative 5. Emphasis in this approach is on *the future,* especially the direction in which current research is heading. Issues taken up might be, Where can we extend our present knowledge? What are the new and emerging areas of study? Are there major hurdles or limitations (other than funding) that must be overcome in order that progress be made? What problems have not yet been rigorously studied? Do we need to improve our research methods, consider new forms of data, expand our analyses, reinterpret previous results? Which recent studies have been more significant in pointing the way forward?

Obviously, these approaches overlap in a number of areas. Most reviews, in fact, combine aspects from two or more of them, but tend to adopt one as their principal orientation toward the subject. It usually helps, therefore, to choose an approach first and then begin to frame your outline around it.

A good way to build your outline is by writing down themes or perceptions you wish to offer—major points or ideas that seem relevant—and then searching for some logical order among them. As always, experiment, and then experiment some more. Put down more subjects than you might need, rather than less. You may have to tinker with your material (add, delete, rephrase) before an order begins to emerge. You may also find that, once these ideas have been put down and given a degree of organization, a different approach than the one you originally decided upon now seems more appropriate. Don't become frustrated or defeatist; such changes are entirely normal and common. The better writer is flexible, adaptable to the material.

How to Begin

The introduction to a review sets up a basic plan for the rest of the article. It identifies the subject and its importance and outlines how it will be treated, what type of perspective is being applied. For this reason, it is worth dwelling on how to begin.

Begin large, at least fairly so. Use general terms to define your topic. This is one of the few places in technical writing where you can offer a broad, open-armed invitation to your readers:

> Explosive taxonomic radiations are very useful to the study of species formation. The extraordinary biological diversity of these systems is seen to evolve through multiple cladogenic events. (for a review on speciation in rapidly diverging systems, with lessons from a specific African lake)

> To survive, living organisms must be able to adapt to their natural environment. Nowhere on earth is this simple evolutionary principle more tested than in high-temperature waters. (on the subject of how thermophilic proteins deal with extreme temperatures)

> Our ability to perceive the dynamics of nature is ultimately limited by the resolution of our instruments. The history of optical instrumentation shows a remarkable advance in recent decades and places us at a new frontier, where ultrafast pulse generation by lasers is key. (on methods and results in the generation of ultrashort optical pulses)

It is perfectly acceptable, in other words, to open with a bit of philosophy, even drama. Different editors, of course, will allow this to different degrees: some may want you to be less grand, more focused on your specific topic:

> The large genomes of mammalian cells are vulnerable to an array of DNA-damaging agents. This situation requires that constant excision and replacement of damaged nucleotide residues take place by DNA repair pathways, in order that potential mutagenic and cytotoxic accidents can be minimized.

Either way, it's good to create interest. Keep your first sentence short, therefore, and follow it up with a longer one that defines, or begins to define, your real subject. You might consider using questions to begin—questions, of course, that you fully intend to answer later on:

> What do we know about the origin of coalbed gases? Where has this knowledge come from, and what are its implications for future energy resources?

Your main effort in the introductory section, however, should be to orient your reader as to the significance of your topic and the direction you're going to take in the review:

> Because of their exotic electromagnetic properties, hole-doped manganese oxides have stimulated considerable scientific and technological interest in recent years . . . The unusual properties of these manganites challenge our current understanding of transition metal oxides and define a fundamental research problem involving study of the interplay between such factors as charge, spin, and orbital degrees of freedom . . . Until recently, there was strong agreement regarding how ferromagnetic states are stabilized in these substances. However, new experimental results suggest that more complex ideas are now needed to explain some of the main properties of these oxides. (Moreo, Yunoki, and Dagotto 1999, 2034)[8]

8. Sentences here have been excerpted and slightly rewritten.

Notice how these sentences develop the subject. First, they set up reasons for interest (aided by such terms as "exotic," "unusual"). Next, the nature of current research is described (this is followed by an explanatory paragraph in the original). Last, the main thrust of the article is revealed; we see what type of perspective the authors intend to add to the subject.

Once you've assembled such an introduction, the rest of the article should begin to fall into place. But—and this is important—be open to changing things as you go along. Again, the process of writing is one of experiment and discovery. You may well find that, in working through your different sections, new ideas emerge, even about the type of perspective you want to give your material, for example, what are the most important limitations to current work, what directions might future research take, and so forth. Since one of your main efforts in writing the review is to survey a given topic and to say what it means, it's essential to be receptive to what the material can tell you.

BOOK REVIEWS

The triumph of the journal in science has not eliminated the importance of books. These still occupy a crucial position. Indeed, there are many kinds to consider: monographs, treatises, edited collections of research papers, transactions volumes, conference proceedings, reference works, not to mention books for more general audiences (many kinds here, from science writing to biography and history of science). Depending on the journal, all of these types may be seen as worthy of review, or just a few. Either way, book reviews now form a regular contribution to most premier journals. And the scientists who write them, and who write them well, provide a crucial service by adopting a certain mantle—they are judges of worth.

Book reviews are commissioned, like review articles. Editors often choose books for review and then assign them to individuals they believe are qualified in the particular field. But there is almost always room for outside suggestion. If you have a work you'd like to review for a specific journal, contact the editor and offer your services: give the title of the book, say why you think it's important to review, and state your qualifications for reviewing it. If your first choice of journal isn't interested, try another. Keep in mind that periodicals are constantly in search of good reviews—these types of articles are widely read and much appreciated by a journal's audience.

So saying, let me make a plea, a necessary one. When you write a book review, please write about the *book*—what's in it, how it's organized, how well it's written, how accurate and complete its information is, how helpful or useful it might be and to whom. Does this seem obvious? But many, many reviews succumb to the temptation, here acknowledged, to wander off into other areas: issues (that the book raises), opinions (of the reviewer), debates, anecdotes, personal experiences, quotations, pet peeves, and so forth. Some of these ingredients are fine, indeed worthy, but *only* when they are woven into a discussion of the work at hand—otherwise, they are little more than pontification. The book is your true subject, first and last.

Book reviews may seem open, casual, and largely nontechnical in nature. Yet an effective review is actually a very tight-knit, even rigorous structure. It provides much information and critical evaluation in the space of a page or less, and may even do so in an entertaining way. Begin by briefly describing the subject of the book:

> The Karoo is an arid and semi-arid region that takes up as much as a third of South Africa and extends north and west into Namibia. It includes large treeless areas, desert, arid and moist savannahs, grassland and, in the most favorable spots, patches of forest. (Bainbridge 2001, 139; a review on a book dealing with ecological processes and patterns in the Karoo)

> Ideas regarding the origins of life have changed a great deal in the past two decades, as an ever greater array of disciplines have begun to tackle the problem. The subject no longer belongs to biology, but has been taken up by researchers in fields as seemingly disparate as astronomy, paleontology, and molecular genetics, with varied but often exciting results. (on a book about the molecular origins of life)

> In the dead of summer 1620, Cornelius Drebbel offered to astound the English royal court by generating a room full of wintry chill. This he did, as we know from the writings of Francis Bacon, but how? We learn the probable answer to this in the opening paragraphs of George Slater's *Conquering Cold: Centuries of Success.*

You can use humor:

> A new textbook on evolutionary biology is likely to find the existing landscape crowded and full of impending extinctions. In order to survive, a text must either fill a specific niche or else range widely enough to cover

> all necessary ground. Remarkably, J. R. Harrison's *Evolution on Earth* is able to do both these things, and thus should provide teachers with the type of work that has long been needed in the field.

Each of these openings sets up a discussion of the volume at hand. What comes next is a description of the book's contents, followed by (or perhaps merged with) judgments about its accuracy, organization, utility, and illustrations. This is a simple formula for writing a review—subject, contents, judgment—but it works, and is easy enough to follow.

Using models here is just as important as it is elsewhere in scientific writing. Keep a file of reviews you like. Read through them before writing your own. In studying their style and content, you may note some of the following points:

- A book review is different, and more interesting, than a book report—which is dry and boring ("This text is the second in a series on soil science . . . Chapter 1 discusses . . . Chapter 2 is about . . . Chapters 3 and 4 take up . . . I didn't like this book, because . . .").
- When being strongly critical of a book, it's best to offer an example or two from the text (evidence) to support your judgment.
- Quotations, well-chosen, provide an excellent taste of a work under review (another form of evidence).
- The pronoun "I" is not usually appropriate to book reviews in science—you are supposed to be giving professional assessments, speaking with a voice of authority for the field, not expounding personal opinions. For these same reasons, the so-called editorial "we" is suitable to use only when you are quite sure you are speaking for most of your colleagues or for the journal.
- Some discussion should be given to any visual materials in a work, that is, figures, photographs, maps, and so on. What is their quality? How helpful and relevant are they? How well chosen?
- A book with too many flaws, about which you have only (or almost only) negative things to say, is probably not worth reviewing. Best allow it to lapse into obscurity on its own demerits. An exception to this is when such a work represents an important position in an area of ongoing scientific debate—here you need to discuss the issues perhaps as much, or even more, than the book itself—or perhaps when the book is written by an especially famous person. Make sure you check with the editor before starting on such a review.
- Remember that no book is ever perfect. The best reviews strive for balance. Impossible standards for judgment serve no one well. Ev-

ery book that's worth reviewing has positive aspects—don't overlook mention of these.

Finally, be cognizant of your potential influence as a reviewer. This influence is real and not to be taken lightly. In providing (let us hope) a professional evaluation, you are effectively giving professional advice. Negative or positive reviews can strongly affect book sales and levels of readership. A highly negative review, on the one hand, is tantamount to don't-buy/don't-read advice and runs the risk of offending both the author and the publisher, at least temporarily. On the other hand, a poor book on an important topic is tantamount to a wasted opportunity for everyone in the field, and this deserves to be pointed out.

None of this, in other words, is a reason to soft-pedal any strong criticism you feel is justified—and that you can justify in your review. It *does* mean that, if pressed, you can back up whatever you say with reason or evidence (or both), that any facts you call upon are truly accurate, and—just as important—that your writing is thoughtful, not glib (criticism for the sake of mere cleverness or for the appearance of superiority does a disservice to everyone). In the end, book reviews are really complex documents of diplomacy. Weigh your words, so they might deserve the authority they will likely carry.

DEBATE/CRITIQUE

Debate is absolutely essential to science. It defines a process by which research results are often confirmed or qualified and new problem areas are outlined. Debate is also how the profession acts to control the quality of its work. Many journals therefore carry a section devoted to critical discussion of previous articles. Typically, this will include one or more critiques of a recent paper and a response from the author.

Most scientists, as routine readers of the literature, are aware of both how necessary and how emotional such exchanges can be. Some instances are congenial, even friendly. Some are almost matter-of-fact or may have a coating of frost. But others are acrimonious. Fur can fly over the use of such terms as "erroneous," "misleading," "invalid," or "specious" when applied to one's lifework. Indeed, there exists an entire vocabulary (largely adjectival) devoted to scientific criticism, and researchers are well-advised to learn how to use it (and withstand it) judiciously. Full discussion of this vocabulary merits a book of its own.

In every case, if you are the critic, consider both the purpose and the effect of the language you employ. Consider what it would be like to be on the receiving end. Your aim in offering critique is to help improve the science at hand, not to cast doubt on the personal credentials or abilities of individuals (except, of course, for cases of fraud, which need to be taken up in a wholly different way). Note the following alternatives:

> In their recent paper on early hominid migration, Leibnitz et al. arrive at conclusions that are spurious and unsupported by any available data. Their interpretation of such migration as a slow, intermittent process runs counter to the results of nearly all current research in the field.

> Leibnitz et al. state their conclusion that early hominid migration was a slow, intermittent process, not sudden and rapid, as other researchers have maintained. We believe this interpretation is flawed for the following reasons.

Here's another example:

> The data provided by Lowell to suggest that clathrates exist near the surface of Mars are erroneous and misleading.

> Lowell offers the intriguing hypothesis that clathrates exist near the surface of Mars. We question this conclusion and the data on which it is based.

And, finally, one more:

> In their study on the eradication of *S. gallinarum,* Klinsmann et al. employed the methods of Barnes (1985), now widely acknowledged to be outdated and error-prone (see Johannson and Walters 1996). Such methods must be seen to invalidate or cast serious doubt upon the results of their study.

> Klinsmann et al. employed the methods of Barnes (1985) in their recent study on the eradication of *S. gallinarum,* an important topic in epidemiological science. I would like to comment on these methods and their possible effects, specifically in view of the critique offered by Johannson and Walters (1996).

In each case, both sentences achieve the same basic goal. This goal is to state the terms of challenge and set the borders and conditions of debate. The difference, of course, is that the second example avoids the implication that Leibnitz et al. are Neanderthals and shouldn't be allowed to practice science beyond the kindergarten level (thinking this is one thing; using words to convey its effect in print is very much another).

Professionalism in this context means keeping a cool head and a restrained voice. That is how you will best serve the field, the research at hand, and (not least) yourself. Like it or not, the words you use will inevitably project an image of yourself as a scientific professional: by keeping cool, you will appear more authoritative and worth listening to. Therefore, consider every use of a critical term carefully, for its effect, and back it up by hard evidence. Keep in mind that too many negative comments will cast doubt on your motives. The appearance of a personal attack will invalidate most of what you have to say. Thus, avoid writing anything in final form when in the throes of anger or some equally intense emotion (disgust, envy, contempt, love, etc.). Always remember that whatever you write, if published, will be there forever, for all to see. Frame your comments so that they focus on the subject matter, not on the "poor choices" made by its authors.

How to do this? Simple: first, jot down your major objections or concerns in the order they come to you, as you read and reread the particular article. Second, look these points over and consider which of them are the most important—you will need to be succinct, as editors demand this. Third, put these points in some order, give them each an introductory sentence ("Leibnitz et al. further suggest that . . . However, this contradicts recent work by . . .") and fill them out with evidence. Your point-by-point discussion can follow the organization of the original paper itself or it can develop its own logic of challenge. Either way, be brief and to the point.

Similarly, if you are on the other side, respond to any such criticism in a like manner, that is, in a professional tone:

> Anderson and Weiss raise some important points in their comment on our paper regarding early hominid migration patterns. However, their criticisms overlook several key factors.

> In questioning the data and conclusions of my recent paper on Martian clathrates, Ardau et al. make the following problematic assumptions.

> While it is true that the methods of Barnes (1985) have been questioned in recent years, many researchers have continued to employ them and, in fact, may be seen to have confirmed their validity.

Answer each point in turn, in the same order it was presented. Be concise and keep your tone even and controlled. Counterattacks ("Anderson and Weiss appear unable to understand the significant points of our paper") will damage your credibility and, in any case, may be rejected or amended by the editor (his or her credibility is also involved). Eloquence in this context of conflict resides in the appearance of intellectual civility and dedication.

Such, in fact, has always been the case. Let us reach back a bit, to an earlier episode of debate, surely one of the most acrimonious and protracted in the history of science.

"Mr. Huxley, was it through your grandfather or your grandmother that you claim descent from a monkey?" So spoke Bishop William Wilberforce before a large public gathering, composed of scientists and laypeople, during an 1860 symposium on evolution held by the British Association for the Advancement of Science. Huxley's reply, you may imagine, is worth quoting:

> If there were an ancestor whom I should feel shame in recalling, it would be not an ape but a man—a man of restless and versatile intellect who, not content with success in his own sphere of activity, plunges into scientific questions with which he has no real acquaintance, only to obscure them by aimless rhetoric and distract the attention of his hearers from the real point at issue by eloquent digressions and skilled appeals to prejudice. (quoted in Huxley 1903, 1:272)

In the midst of so public a forum, with everything to gain or to lose, Thomas Huxley was not above complimenting his antagonist, even while destroying him with tact and aplomb.

CHAPTER NINE

Graphics and Their Place

VISUAL LANGUAGE: SEPARATE BUT EQUAL

Modern science relies deeply upon illustration—graphs, charts, drawings, photographs, maps, models, and other forms. Technical knowledge today is inseparable from visual presentation, from its specific powers to order and convey information. Scientists, moreover, appreciate excellent graphics. Illustrations that offer data with clarity and elegance are a unique type of achievement—creative, efficient, even a source of delight.

Scientific illustration has an extremely rich and venerable history, reaching back to the written works of ancient Egypt, Greece, China, India, and medieval Islamic civilizations. Alloys between science, art, and draftsmanship, forged in the European Renaissance and after, are still evident in many aspects of contemporary image making—in drawings of specimens, attention paid to form and balance, three-dimensional effects, uses of color. Many of the most influential works in the history of science—Galileo's *Sidereus nuncius* or Vesalius's *De humani corporis fabrica,* for example—have been books of pictures as well as text.

That said, it should be stressed that the visual dimension to science forms a language all its own, a kind of pictorial rhetoric, if you will. By this I mean that graphics are often much more than a mere handmaiden to writing. They don't just restate the data or reduce the need for prose, but offer a kind of separate "text" for reading and interpretation. To assure yourself of this, take any well-illustrated article, copy the figures, and assemble them in order of appearance. You will find that they tell their own story, in some manner parallel to that of the writing, but in other ways different, enriching, though also with notable gaps.

Illustrations serve a variety of functions. Charts summarize data and make comparisons. Graphs provide analysis by revealing patterns, relationships, or possible correlations. Images, meanwhile, offer different kinds of evidence, explain and explore information, demonstrate specific points, represent concepts or theories. All in all, this is an impressive array of service—and it certainly helps point up (and justify) why scientists often browse through articles by reading the abstract and looking at the illustrations.

Perhaps most fundamental of all, however, visual discourse adds variety for the eye and enhanced appeal for the mind. Does this seem trivial? It shouldn't: the psychology of reading is not a little complex. The living brain very much appreciates intelligence expressed in different forms.

SPECIFICITY AND CHANGE

As a scientist-author, fresh on the trail of publication, it's a good idea to become intimately familiar with graphic elements used in your field. This may seem obvious. But there are two factors that raise it beyond mere common sense.

First, many visuals are highly field specific. Even graphs, maps, or other presumably standard figures can change considerably in form and style, as well as content, between disciplines. Moreover, there is the "journal effect" to consider: even within your own field, different periodicals have their own demands for how they want articles to look, just as they have standards for written copy.

A second reason to get familiar with graphics in your area is that many types of visuals are undergoing change, due mainly to the advent of digital technology, and will continue to evolve as this technology does. Indeed, the digital age has introduced a fertile array of new visual possibilities: satellite imagery, three-dimensional modeling, ultrasound technology, tomography, magnetic resonance imaging, various types of electron microscopy, and a dozen others. Science is replete with new powers of vision, new means of making the informed eye the instrument of study, analysis, and discovery. Color, too, has found a new and expanding role in scientific imagery and continues to revise older forms. Those who grew up in science before the 1980s can testify to a former universe of black and white, where the tones of text and image were identical. Since then, a great separation has taken place, with color now abloom in many fields, being much more easily and routinely achieved (though not cheaply, by any standard). More than ever, pictures today often demand and reward attentive study in themselves.

Becoming familiar with the graphics of your field is therefore essential—but it is not enough, in and of itself. The new sophistication in imagery depends upon software, which is also often designed for individual fields. Certainly, a number of generic programs exist to help scientists create graphs, charts, and tables. Most major software vendors offer programs that cover basic graphing capabilities, useful in many circumstances when simple plots will do (Microsoft Excel is one of the most widely used). For more complex demands, there are programs designed specifically for scientific uses (Sigma Plot and Axum are two popular choices). Beyond these general-purpose programs, software is likely to be specialized. Fields as diverse as molecular genetics, petroleum geology, climatology, and mechanical engineering now each have a large spectrum of dedicated graphics programs to use with particular types of data. This places new demands on the scientist, who often must learn at least some of these programs.

Finally, new visual possibilities are attached to Internet communication. Still very young, Internet science promises new types of nonwritten expression: real-time animation, interactive modeling, use of video and audio. In some areas, these capabilities have already been taken up—online journals in medicine and chemistry, for example, have included animated graphics, video, and interactive visuals of various types. Change is in the wind for scientific illustration, and on the ground.

All of which highlights, again, the need for the individual scientist to learn what is out there. Becoming an effective author means becoming literate in these forms of communication—learning how to read them, how to recognize good *and* bad examples, when and where to imitate the best.

CHOOSING MODELS: A HELP HERE, TOO

As with writing, you can learn a lot about producing good graphics by studying the admirable work done by others—and the opposite. Not only will this provide you with guiding examples to emulate and avoid, it will also help sharpen your critical faculty about what goes into such an image, what makes it effective, easily deciphered, informative, attractive. You can collect entire articles or only one or two illustrations from a paper. You might want to make a fairly large collection at first and then whittle it down to the very best: the exercise of doing this will decidedly focus your attention on details of quality. You might try gathering several nice examples of a single type of image, compare them side-by-side, and choose from there. Any method that helps you scrutinize and evaluate the images of your field is valuable. Find out what frustrates or irritates you (Too much type? Text too small? Figure too busy? Poor labeling?). The best questions to ask, as always, are, Is this something I wish I had done? or What would I do to make this better?

For every one of your choices, or for any graphic you find particularly worthy (or the opposite), stop and ask yourself what it is that seems especially good (or poor) to you. The more you come to understand your own preferences, the more you can use them consciously in preparing your own articles. Here are a few categories to help you judge individual illustrations and analyze your impressions.

- *Neatness.* Is the image clean and sharp; does it invite attention, or repel it?
- *Readability.* Can the eye move over it and pick up information, either quickly or with concentrated attention, or is there too much confusion, too much data (a "crammed" feel), a lack of integration?
- *Use of Type.* Is the font easily readable, large enough; is it placed well, or does it invade and distract; is there too much of it (a common error)? Is there a proper hierarchy among type sizes—do the largest words refer to the most important items, the smallest to the least important?
- *Size.* Is the image too small to be fully visible; do you find yourself drawing the page nearer to your face? On the other hand, does it fill the rectangular space to the edges, as it should, or does it have too much white space?
- *Aesthetics.* Is the image balanced, or does it seem lopsided? What aspects or portions draw the eye most, and are these significant in terms of content? Are the thickest lines on the graph the most important; do the patterns in a bar chart highlight the differences you want to show?
- *Use of Color.* Does it help distinguish content, increase readability; are colors appropriately distributed (e.g. blue for depth, red for height); is text minimized yet visible; is the legend (if needed to differentiate the meaning of each color) simple and easy to use? Are colors consistently coded among different images?

- *Consistency.* Do similar images (maps, charts, graphs, photographs, etc.) carry the same stylistic scheme in terms of line width, type font and size, labeling, scales, and so forth?
- *Room for Improvement.* Are there any changes you yourself might make to improve the quality of this image?

EXPERIMENT, EXPERIMENT

I said earlier that writing involves experimentation—trial and error, revision, working things out. This is equally true for figures. It is true even for seemingly simple visuals, such as charts or graphs. We often construct these from our data as tools to help us in our analysis, to make comparisons, look for trends, discover relationships. When we initially draft our illustrations, we are frequently performing what are akin to visual trials—recasting our information in new forms to see if something important and unforeseen steps into the light.

The experienced scientist-author knows, moreover, that it can be very helpful to take a single data set and graph it in different ways. This means trying out distinct analytical versions in order to discover what form offers the most effective, meaningful presentation. Today this can often be done for charts and graphs, in particular, with a few clicks of a computer mouse: histograms can be changed to line graphs, pie charts, dot graphs, and other forms, with or without labels, error bars, means, and averages. Current software gives you the power to play with data in productive ways. As with text, the process of creating any specific image can reveal new aspects or relationships previously unnoticed.

So it is entirely normal—indeed it should be expected—that you'll often create more illustrations than you need, and that you'll need to revise the ones you keep. Sometimes a particular graphic, one you may have worked hard on, will be unnecessary; its message can be stated in a sentence or two of text—if so, please (for the sake of scientists everywhere) delete it, grieve briefly, and move on.

Experimenting with graphics also means making decisions about appearance. In a good scientific illustration, everything visual qualifies as content. Choices therefore need to be made about such things as, how much text should be used; what type font and size are best; what shadings or colors are appropriate; what kind of scale should be used; how far should each axis extend; how wide should the bars or lines be on a chart or graph; what patterns should be used; is a legend needed and where should it go; and so on. For more complex figures, the decisions are likely to be even more numerous.

Luckily, the relevant answers to such questions are far from open-ended. There is much standardization in scientific imagery, and your models will help guide your choices. No example, however, is ever a final template. There will also be factors (let us hope) individual and original to your work, and these will require that you adapt particular graphical forms to your specific case. This may involve changing scales, altering colors used in a model, graphing an added variable.

Think of each figure, therefore, as a draft in the beginning, a kind of visual audition. No time spent trying things out intelligently is ever wasted. It is sometimes necessary to discover what graphical forms might work best with your data, which types of illustrations you are most comfortable using. After all, this too is an area where authors develop a certain style, however subtly expressed. Don't berate yourself for any dead ends; be thankful when you find and overcome them. Experimentation, in both writing and illustration, is one of the most crucial processes in learning how to communicate well.

Many articles need (but often don't have) an introductory illustration to help orient the reader. Could your document benefit from such a graphic? The answer might be yes, no, or maybe, but it's usually worth asking and thinking about the question. An introductory graphic can serve many of the same purposes as the introduction itself—it can provide setting, such as can be done with a map, large-scale model, and so forth; it can offer essential background, for example, in the form of a flow chart that shows a sequence of relationships, a schematic diagram of apparatus, a time-based graph or chart outlining the relationship under study. Again, a guiding principle is to think of what might help usher your reader into the domain of your investigation. What would you use in an oral presentation to do this?

Try to make sure that any diagrams and drawings are relatively pleasant to look at. Use a type style that can be read very easily. Lowercase lettering is usually more pleasant to read than all uppercase. Many journals prefer a sans serif font for illustrations: Arial and Helvetica are common choices. Others prefer serif styles or are nonspecific. Check this out *before* you create your images—spare yourself (or your artist) the agonies of unnecessary revision.

As with text, different journals will have different specifications for many aspects of the illustrations they agree to publish. At a general level, this is likely to involve such aspects as software format (preferred programs), sizing, method of delivery (whether on disk, via the Internet, or as printouts). Some journals, however, are quite specific about the detailed form your graphics should take, for example, type style and size, line thickness, use of borders and arrows, labeling (where and when), scale bars, and more. Be aware: certain periodicals spell all this out in their instructions to authors, but some do not. Some demand consistency of stylistic detail among articles, and some are less exacting. Thus, you really do need to take a close look at published examples in your journal of choice before spending the time to design and create your images.

SOME NECESSARY POINTERS

Keep text on your figures to a minimum. Use it mainly for labeling, not for explanation (leave this to the caption and main text). Also, be consistent from one figure to the next in the fonts you use, numbering and lettering style, and other such aspects.

Avoid any overly fanciful or arcane fonts for your images. They will distract the viewer (while drawing attention to yourself as the possible embodiment of bad or eccentric taste).

When needed, use different font sizes to indicate different levels of importance. Be consistent about your sizes and font styles from one graphic to the next. Alternatively, consider using boldface or italics as highlights or as a means to add a dash of visual interest. Be aware, however: too much variation is distracting, so keep your visual hierarchy simple, clean, efficient.

Design your figures so that they extend nearly to the edges of the frame: do not waste space. This will allow you to make each graphic as large as possible, which is very important, because—

Images in science are nearly always reduced considerably for printing, commonly 50% to 75% (one-half to one-quarter original size) or more. Therefore, *anticipate the effect of reduction* (not doing so is one of the most common errors made in scientific illustrations). Use a copy machine (or computer reduction, if the figure is a digital file) to reduce your images and test this out: make sure all text, lines, and details hold up. Another test is to take images that have been published and enlarge them to roughly the size of your own illustrations, then compare type size, line thickness, and so on.

Figures created for slide and computer presentations are almost *never* suitable for print journals, reports, or books. These presentation images must be translated, or even redone, for hard-copy debut. Frequent changes required include less text, smaller font sizes, thinner lines, and removal of color (simple gray-scaling very often isn't enough). In general, presentation images are similar to cartoon versions of what should appear in print.

Make sure all digital illustrations, whether originals or scans, have a resolution of 300 dpi (dots per inch) or higher. This is the absolute minimum required for printing; 600 dpi is usually better. Provide high-quality laser printouts of every figure.

Use a simple, clear system for naming digital files of your artwork. Common schemes include your last name (lead author) + figure number (Montgomery.2); one or two keywords + figure number (Apoptosis.2); and simple abbreviation of keywords + figure number (CrysRNA.2).

True elegance in science resides in simplicity and restraint.

EXAMPLES

In the pages that follow, I've tried to present a series of good and bad examples of illustrations taken from the published literature. In each case, a brief commentary, and possibly a question or two, are given in order to help you evaluate the relevant image and thus further your own critical ventures.

Charts and graphs are particularly well represented in figures 9.1–9.10. This is because these graphics are surely the most ubiquitous in all of science. My own unofficial survey of more than 150 journals in 57 fields (from insect physiology to mathematical physics) shows that they occur almost twice as often as any other type of visual. Indeed, in some periodicals, they are the *only* graphical form to be found. Thus, their importance is rather high.

Tables present exact numerical data, whereas charts and graphs take these data and give them a form of visual analysis. Use tables when you need to show specific or precise values; use charts and graphs when you want to find and express meaningful relationships from these numbers. Very rarely will you ever need to show both.

Notice in the figures that follow that, in most cases, the horizontal axis plots an independent variable—the data set that we select—whereas the vertical axis gives the dependent variable—what we measure.

Bar Charts

Bar charts or graphs, particularly histograms, are extremely common. The essence of such charts is usually to make comparisons between data

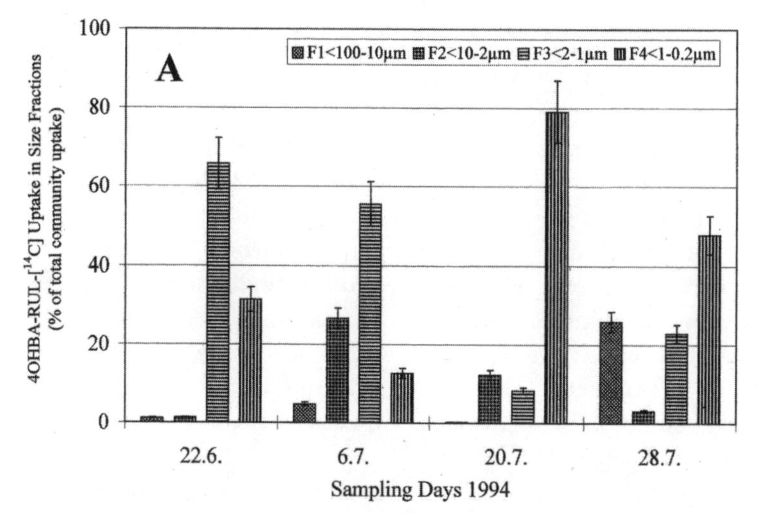

Figure 9.1 Example of a bar chart (from Munster et al. 1999, 118; used with permission of Schweizerbart Publishers, www.schweizerbart.de)

sets. They are thus particularly useful for showing differences, and are less suited for revealing trends or relationships.

Take figure 9.1. This shows the uptake of a particular carbon-bearing chemical by a forest lake, during a series of sampling-day intervals. Several aspects of this chart seem well done. First, the axes are labeled in type large enough to easily read, and a legend is provided. Second, the bars are of adequate thickness (this is not trivial) and show appropriate error ranges. Third, the different data sets are properly separated. Fourth, each data set shows the same order of bar patterns.

What can be done better? The *x*-axis is improperly labeled: shown are sampling *dates* (day, month), not days. The patterns for individual bars are much too similar to be easily distinguished. Note, too, that the legend is too small and cramped. These problems leave the data difficult to decipher and thus interpret. Solutions to any and all of these problems are simple and straightforward.

Now examine figure 9.2. This box-and-whiskers chart is an attempt to show the effect of parasitism on weight gain in two species of butterfly (*Heliothis virescens* and *Trichoplusia ni,* abbreviated "Hv" and "Tn" in the figure). Nearly everything is clearly labeled on this figure; the type is large and welcoming. Bar patterns are easy to distinguish, the legend is neatly done, and the interpretation of the data is clear (parasitism halts growth at an early stage). Moreover, the author has nested the bars for each individual species to highlight the comparisons being made.

Could anything be improved? The whiskers sticking out of the top of each bar could be explained (error bars?). The intervals on the horizontal axis could be a bit wider, allowing for a widening of the bars. d10 could be placed in its rightful interval, rather than in d9—which suggests that perhaps two-day intervals (d2, d4, d6, . . . , d10) might have been a superior choice for sampling. Finally, there is a larger question: would the data be more revealingly shown in a line graph? If the authors' point were to show only gross, overall patterns, then the answer would be no (or rather, not necessarily). But if ideas of developmental progress are at issue—ideas that depend on continuity through time—the answer would be yes, for discontinuous samplings would no longer suffice (why offer snapshots, when the entire movie is available at no extra charge?).

Now consider figure 9.3. This is a simple, vertically oriented chart, showing average measured porosity for petroleum-bearing rocks originally deposited in different settings. The data are averaged from a wide range of samplings and are thus quite generalized; the graph is meant to show only very large-scale comparisons. As a result, it probably doesn't need either the detail given in the vertical scale or the related horizontal ruling. We do need to know what average porosity is measured in, however; what do the numbers along the *y*-axis represent?

The bars are generous in width, but are not separated (as they should be), and the patterns are confusing—are we to assume some relation

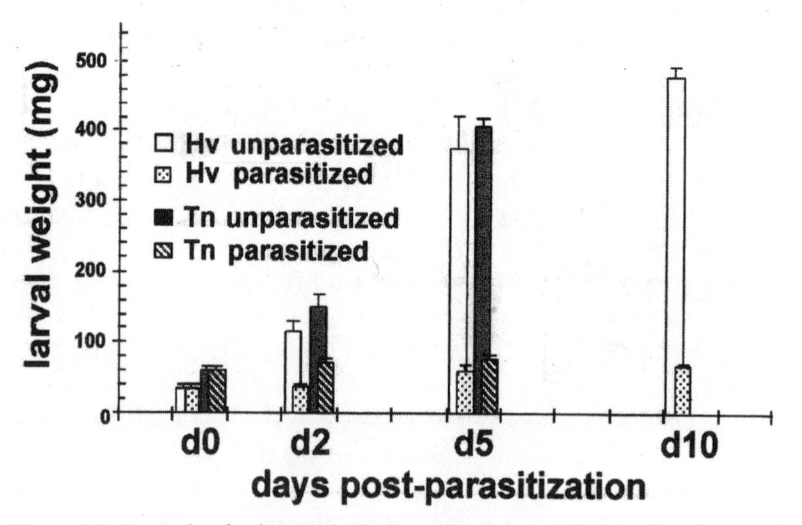

Figure 9.2 Example of a box-and-whiskers chart (reprinted from Cui, Soldevila, and Webb 2000, 1401; copyright 2000 Elsevier Science; used with permission)

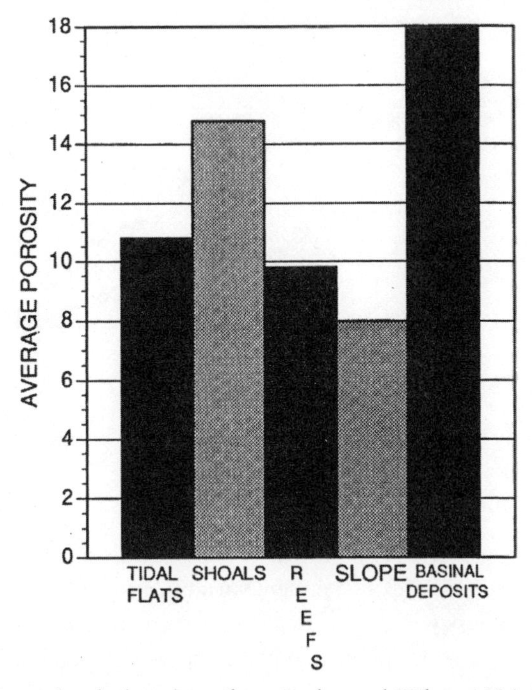

Figure 9.3 Example of a bar chart (from Jordan and Wilson 1994, 148)

between those similar patterns (e.g. shoals and slope)? Confusion is avoided if even a little space is added between each bar and a single pattern used. Note, too, the identifying text along the horizontal axis: again, are we being given clues that "Reefs" (written differently) and "Slope" (in larger type) merit particular attention? Uncertainty here can be averted in several ways, for example, by using different patterns for each bar and constructing a legend; by placing the respective label *above* each bar (which are at different heights) or even *within* it; or, again, separating the bars a little and making the type size a bit smaller and consistent.

Line Graphs

Nonbar graphs come in even more numerous varieties. With graphs you can show trends, correlations, and frequency distributions (various line graphs); rates of change (semilogarithmic graphs); changes in relative difference (area graphs); patterns among discrete random variables; and much more. The essence of a graph, however, in a majority of cases, is to show continuous relationships: data exist in some type of continuum defined by dependence between variables.

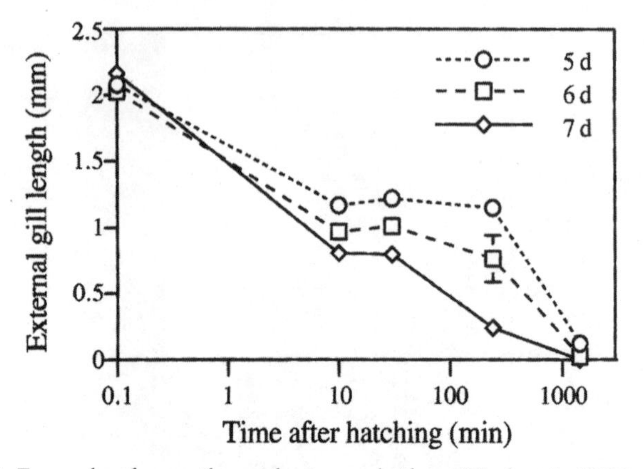

Figure 9.4 Example of a semilogarithmic graph (from Warkentin 2000, 559; used with permission)

Figure 9.4 presents a simple and wholly effective semilogarithmic graph, where time is the independent variable and is given in logarithmic scale. Nothing fancy is needed here; the graph compares changes in external gill length of tadpoles hatched at three different ages (5, 6, and 7 days). The text is clean, minimal, consistent. The axes are neatly labeled. The lines are clear and easily distinguished, and the data points are marked. Note that the logarithmic scale (x-axis) shows actual values (0.1, 1, 10, 100, etc.), not the logarithms (–1, 0, 1, 2, etc.). For much larger or smaller numbers, it is common to write in powers of ten, that is, 10^5, 10^6, 10^7 or 10^{-4}, 10^{-5}, 10^{-6}, and so on.

Compare this graph to figure 9.5, which includes three related plots, showing the concentrations of methane, carbon dioxide, and water in sediments of increasing depth below the sediment-water interface (marked as 0 on the vertical axis). Each graph carries six lines—six data sets—indicated by well-chosen standard symbols. There is some question as to whether the data for methane and carbon dioxide can be meaningfully distinguished and interpreted beyond general trends (which begs the question of why individual data lines are needed). In the graph for water content, meanwhile, this is less important, due to the tight clustering. How might the graph be improved? Here is a case where using color would make eminent sense. Possibly, the plots for methane and carbon dioxide would gain a bit of clarity (and meaning) if the intervals on the horizontal scale were widened. Finally, only one legend is really needed (it's the same for all three graphs).

Deciding how many lines to plot, how many correlations to reveal, is obviously very important and can require experimentation. As figure

9.6 shows, it is sometimes possible to include many lines in an effective manner. Notice here how well-chosen the line patterns are in terms of the eye's ability to differentiate—an aspect that adds both clarity and visual appeal. At the same time, however, this is possible only because the lines do not cross each other very much. Were they to do so, it would be necessary to break the data out into two or three separate graphs.

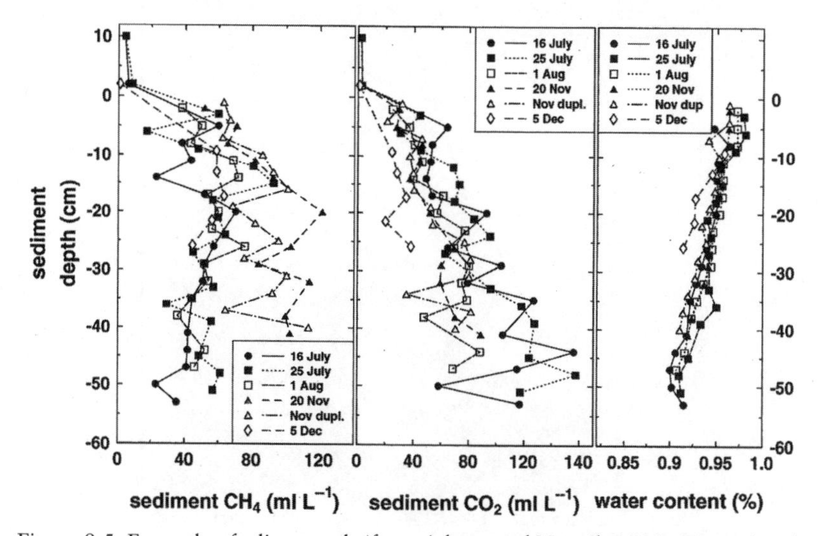

Figure 9.5 Example of a line graph (from Adams and Naguib 1999, 94; used with permission of Schweizerbart Publishers, www.schweizerbart.de)

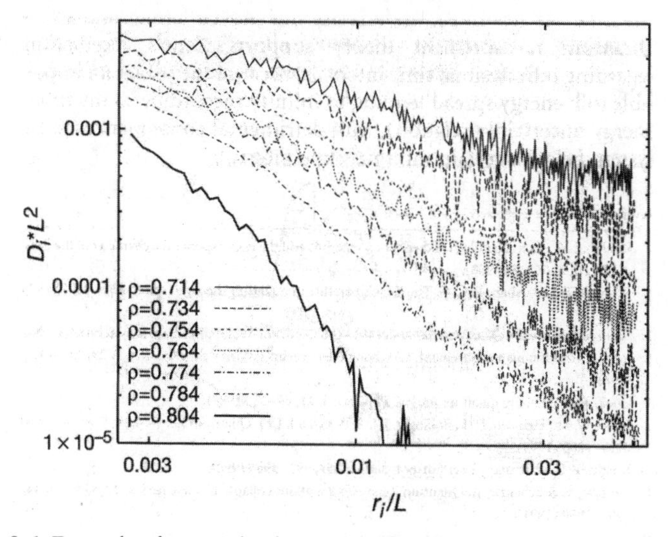

Figure 9.6 Example of a complex line graph (from Santen and Krauth 2000, 550; copyright 2000 *Nature;* used with permission)

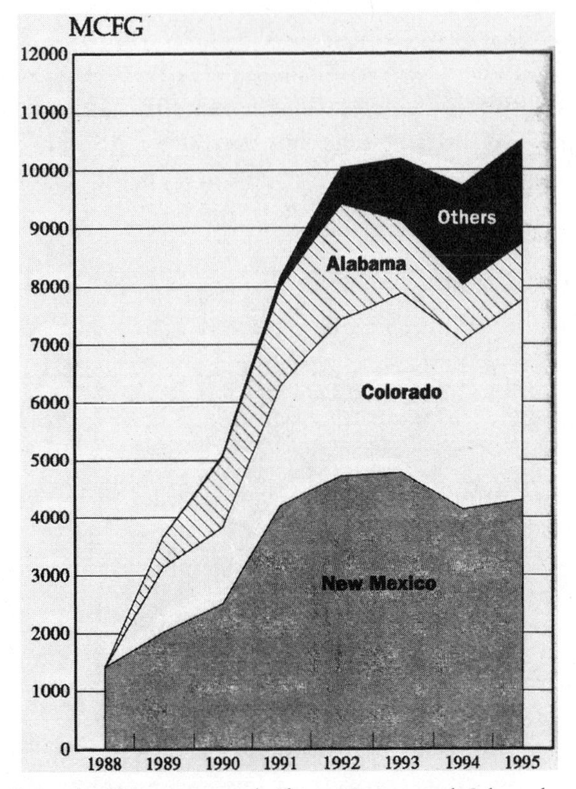

Figure 9.7 Example of an area graph (from Murray and Schwochow 1997, 32; used with permission)

Visually speaking, moreover, not all lines are created equal: note how the thicker, solid plots draw attention and imply importance, while the thinner, more broken ones carry much less psychological weight. Perhaps the only way to improve this example is to make the *y*-axis more consistent in interval labels, that is, 10^{-5} (we don't need the "1 ×"), 10^{-4}, and 10^{-3}.

Another type of line graph, sometimes called an area graph, appears in figure 9.7. This type of data display is used to show both progressive change and comparisons between different data fields. Note how important it is to use different and easily distinguished patterns within each data field. Color is not necessary here, but could be a help if a larger number of fields were plotted. Small improvements might be made: the axes might be better labeled (MCFG might be written out along the *y*-axis, "Million Cubic Feet of Gas"); we might want to extend the data fields through dashes (projected) to the right margin of the graph. On the whole, however, this is an informative and well-done figure.

Maps and Diagrams

Figures 9.8 and 9.9 show maps with a specific problem. Figure 9.8 is meant to compare precipitation levels associated with the 1997–1998 El Niño phenomenon, but the maps are far too small to read and interpret on any intelligent basis. In such cases, the editor of the journal and the authors of the article need to make a decision that favors their readers, not the data alone. If no more space could be allotted these images than is shown, they should have been deleted or else one map selected and shown at larger size. As it is, they provide more in the way of frustration than enlightenment.

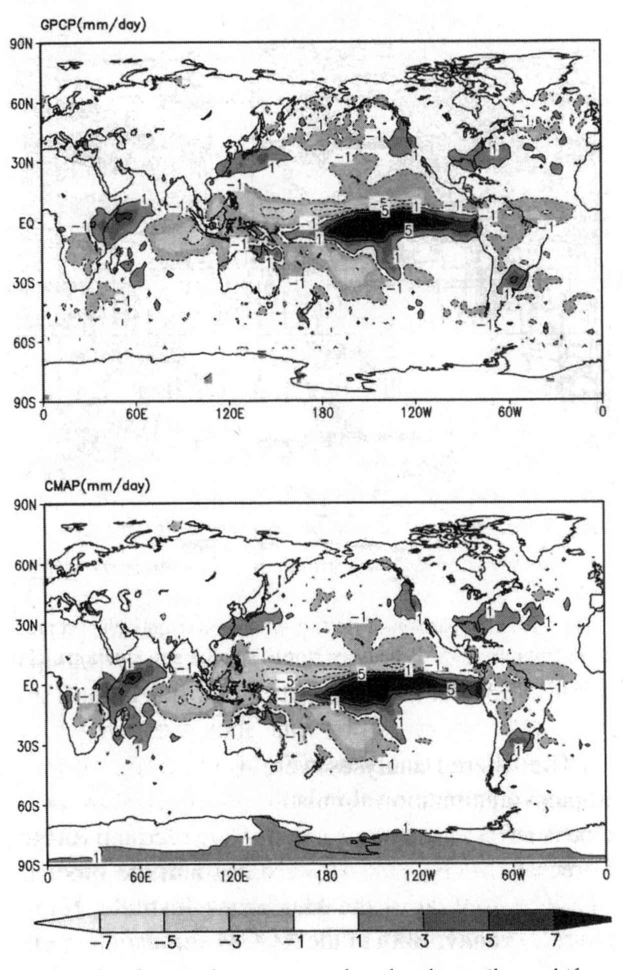

Figure 9.8 Example of maps that are too reduced to be easily read (from Guber et al. 2000, 2641; copyright American Meteorological Society; used with permission)

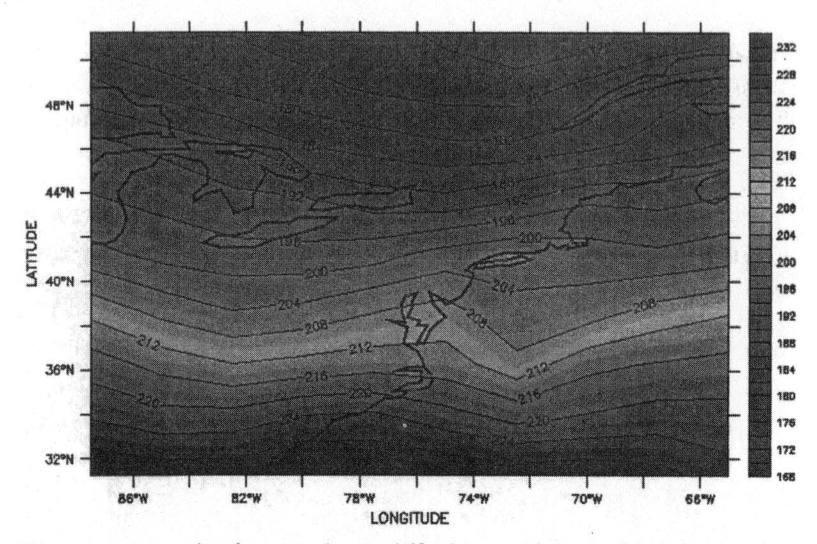

Figure 9.9 Example of a map that is difficult to read due to loss of color (from Bess et al. 2000, 2649; copyright American Meteorological Society; used with permission)

Figure 9.9, meanwhile, offers us another type of disappointment. Images originally drafted in color, for talks or poster presentations for example, do not always translate well into black and white or gray scale. Here, the data would be illegible if the contours weren't numerically labeled; certainly the scale on the right is of little help. The figure would have been clearer if it lacked shading altogether and merely offered its contours against a white background. Were color to be restored to this map, however, its message would be enhanced enormously, well beyond any black-and-white version. The power of color to reveal gradation and to highlight peaks and lows in such a data set is great indeed, and can offer the eye pleasure and the mind definite advantage. Such advantage, to be sure, must be weighed against cost. If color proves too expensive, yet is deeply integral to the meaning, the figure may need to be redrafted, as is the case here.

Which brings us to the example of figure 9.10. Let us admit, up front, that computer-generated presentations are a major advance for scientists. Indeed, they are far more simple, efficient, and (even) fun to create than the old, laborious, hand-drawn figures and typed-out tables we used to produce. But this new ease of creation comes with a price. In a great majority of cases, illustrations generated by computer for slide or other presentations do not make good visuals for published articles—unless revised with hard-copy standards in mind. Briefly put, we are dealing here with two very different media, each with its own distinct needs and limits. A graphic like the top panel of figure 9.10, with its verbiage surrounding and inhabiting the data, plays very well up on the big screen, but looks clumsy and amateurish in print. Much of the writing here is interpretive and belongs in the main body of the article itself. To appear in a high-quality journal, the graphic would need to be revised as shown in the bottom panel of figure 9.10: title removed, interpretive text deleted, axis labels reduced in size, lines thinned.

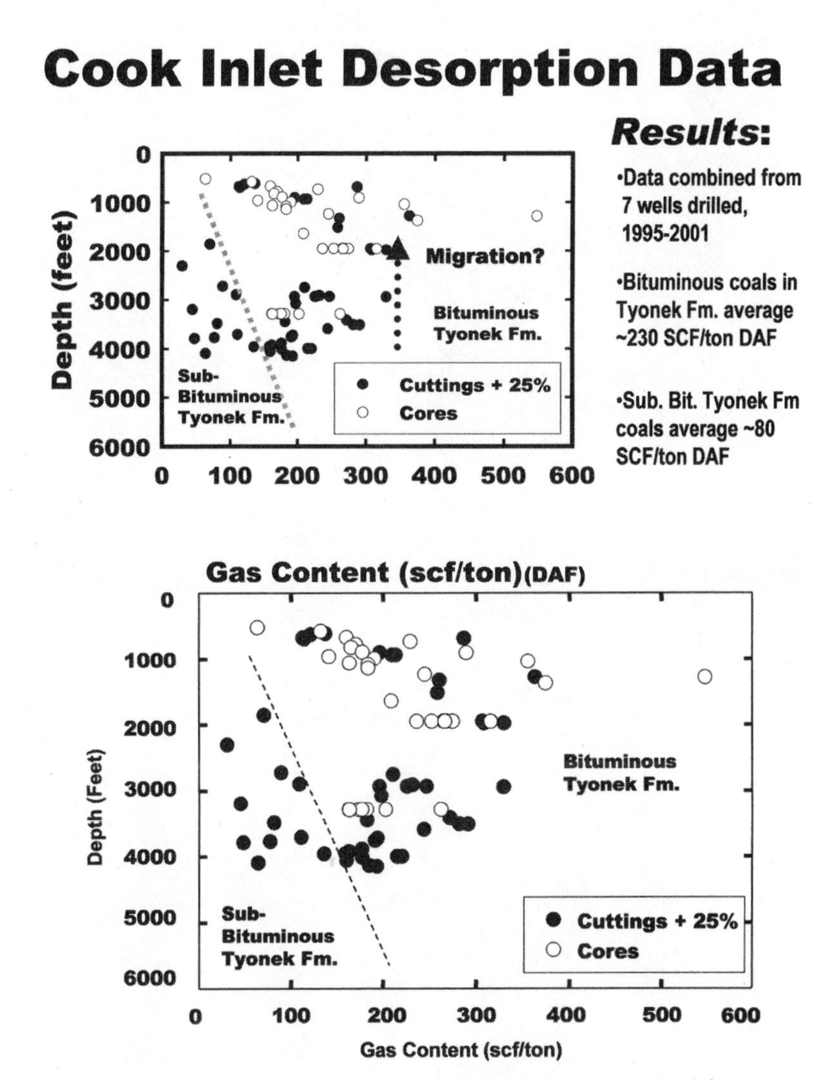

Figure 9.10 Example of computer-generated graphic. *Top*, unedited from use in oral presentation; *bottom*, properly edited for print

Schematic diagrams are quite common in science and, as their name implies, are usually most effective when kept as simple as possible. Figure 9.11 is an illustration of a dipolarization model that shows this very well. The image employs basic shapes and visual elements (circles, dots, arrows, different types and thicknesses of lines). It makes good use of space, being neither crowded nor too open. It includes a minimum of text; in fact, the use of words on the diagram is a good thing, since it gives us valuable orientation and prevents the figure as a whole from becoming a mosaic of single-letter symbols.

Another good example is figure 9.12, which depicts how a particular

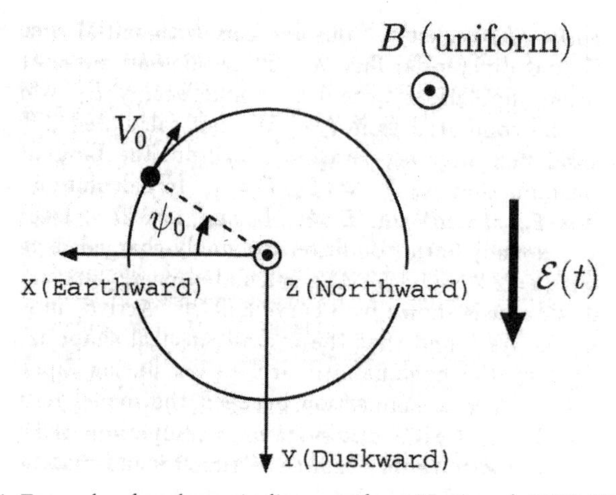

Figure 9.11 Example of a schematic diagram (from Nosé et al. 2000, 23283; copyright 2000 American Geophysical Union; reproduced with permission)

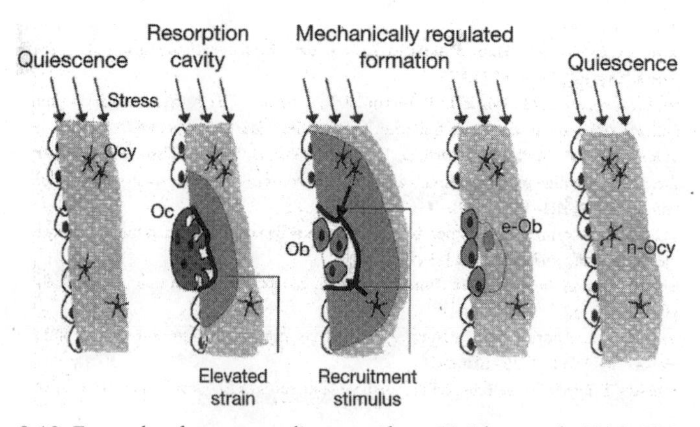

Figure 9.12 Example of a process diagram (from Huiskes et al. 2000, 705; copyright 2000 *Nature;* used with permission)

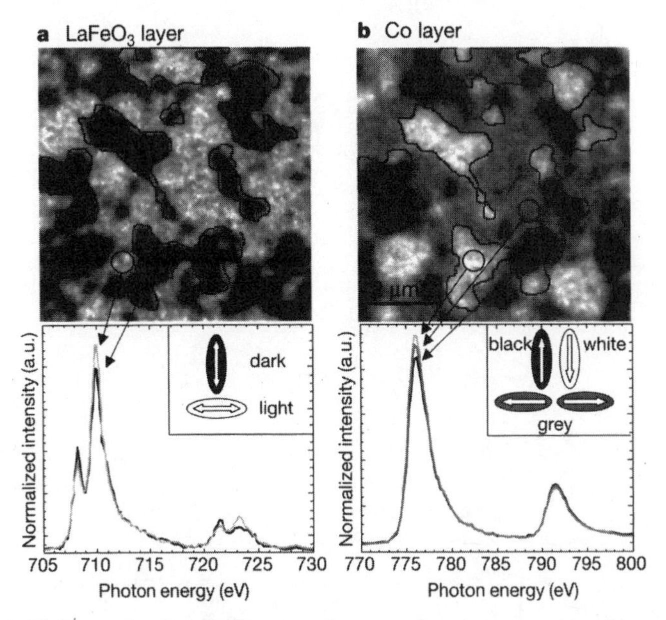

Figure 9.13 Example of a combination diagram, showing two types of images successfully integrated (from Nolting et al. 2000, 767; copyright 2000 *Nature;* used with permission)

type of bone responds to an applied stress. In this case, the diagram shows a cyclic progression of states, involving a temporary weakening of bone mass (through the formation of a cavity and resulting increase in strain) and its repair. A reader can follow the process easily, from left to right. Again, there is minimal text, just enough to identify crucial elements (all abbreviations are explained in the caption). Color is not needed here, since shading can accommodate everything shown. Is the figure perfect? No such phenomenon exists. Some scientists would find value in numbering each of the stages and using these to explain the process in the caption or main text; this would be efficient and effective and would not add overly to the information shown. Others might provide a legend explaining each of the shaded areas, as well as abbreviations.

Combination diagrams, in which two or more types of illustrations are grouped together, have become quite common in many scientific publications. This has led to many fertile blendings of visual information. In figure 9.13, images and spectral graphs are nicely juxtaposed to illustrate behavior of alternating ferromagnetic and antiferromagnetic layers within a given substance. This figure contains a large amount of information, but presents it clearly and even elegantly. Note, for exam-

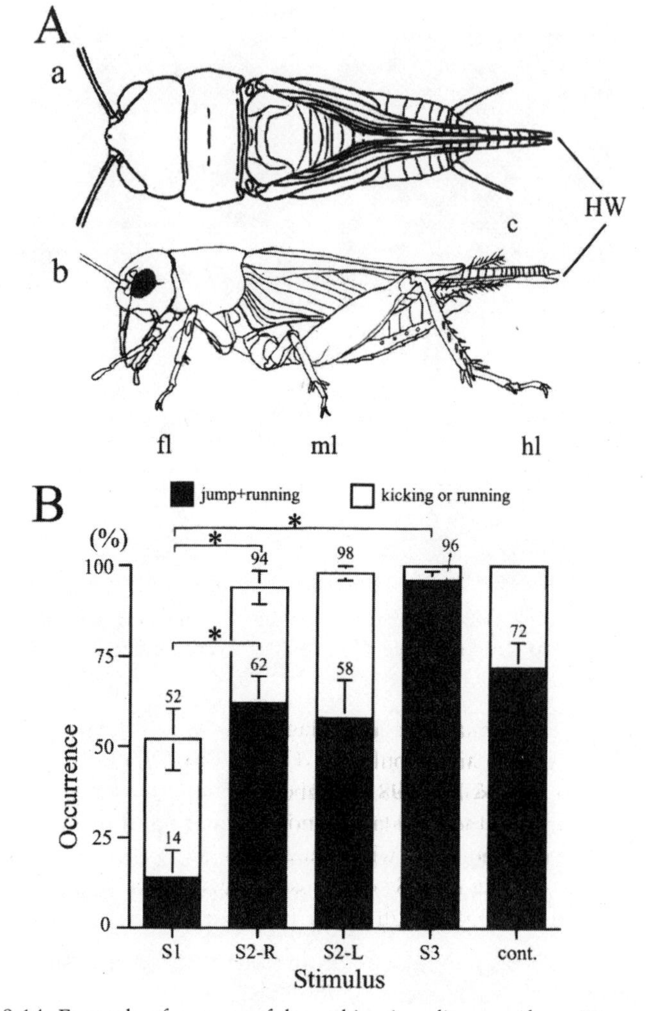

Figure 9.14 Example of a successful combination diagram (from Hiraguchi and Yamaguchi 2000, 1332; copyright 2000 Elsevier Science; used with permission)

ple, how each half of the total is visually balanced, with the legend inserted neatly into the upper right portion of the respective graph (which would otherwise show a lot of white space). Small arrows are used to tie portions of each image to its respective position on the spectral graph. Though the overall size of the figure is small, the text and numbers are large enough and in an appropriate font to be easily legible.

Another example is given in figure 9.14. Here a specimen drawing is combined with a histogram. The drawing indicates the few key anatomical features of the cricket relevant to the experiments, with the most prominent of these (HW) shown in capital letters to emphasize their central importance. The histogram, meanwhile, plots two types of escape behavior for four different kinds of stimuli (plus a control group), each of which was applied to the HW (hindwing tip). At first, everything looks to be in good order. But when we examine the details of the figure, a few problems emerge. Notice that there are two sets of *a* and *b* images, one in uppercase, the other in lowercase, plus a *c* designation on the upper specimen drawing. Probably the lowercase *a* and *b* could be replaced with numbers or deleted altogether. It would also help, to avoid any confusion, to place the legend (black box, open box) at the bottom. There is also enough space on the figure to write out the words for the anatomical features indicated on the drawing, none of which is very long (foreleg, midleg, hindleg, circus, hindwing). This would reduce the burdens on the caption to list and explain a large number of symbols, a consideration worthy to be granted the reader. There is one other problem with this figure that we will leave aside for a moment.

As a final example, I turn to an illustration that reveals how far the combining impulse has been taken in recent times. Figure 9.15, from the premier journal *Cell,* shows six different types of images merged into a single visual conglomerate. These images include a schematic drawing, two line graphs, a scattergram, a histogram, and two tunneling electron micrographs. The caption, not surprisingly, reads like a headline: "*Su*(*H*) Auto-Activation is Required for Normal Mechanoreceptor Function," with the following explanation of each segment (A–G) requiring nearly half a page.

Many editors, I imagine, would go to war over this sort of thing. But the fact is that it has become an established and often-used *modus illustratis* in certain branches of science and serves a valuable function. Not only does it offer more data, often important data, than would be possible if each graphic were given individual space; it also creates something like a miniarticle within the larger paper or report, thereby providing a kind of added expressive dimension to the whole. Obviously this is an approach that can easily be taken too far; visual chaos and confusion are the inevitable result if too many dissimilar graphics are crammed into a single space. Figure 9.15 suggests the limit, though (in my opinion) does not approach it. Such a medley should never be done pell-mell. A clear progression and logic must be established between the individual images. Data dumping is an affront to the reader, no matter if done in images or in writing.

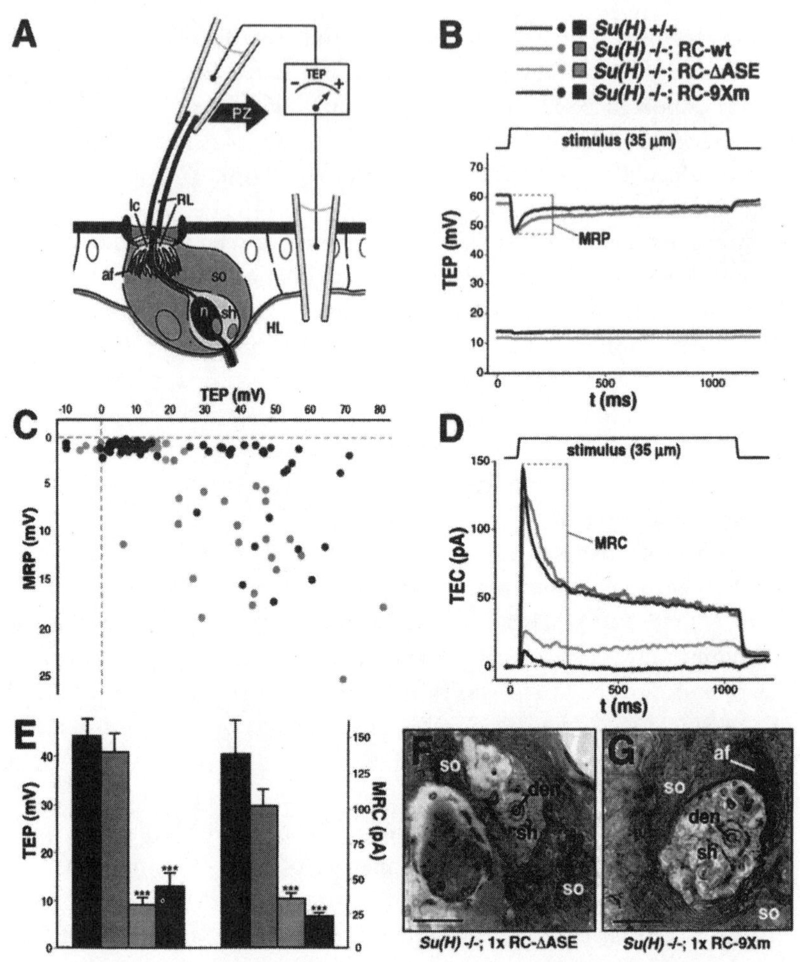

Figure 9.15 Example of a highly complex combination diagram, involving a wide range of image types (from Barolo et al. 2000, 964)

REFERRING TO ILLUSTRATIONS IN YOUR TEXT

Part of the craft (and sometimes art) of good scientific writing is knowing how to integrate graphical materials. Visually, illustrations are embedded in the text, like boulders in a stream, but you want them to be part of the flow as well. Perhaps a better metaphor is the quilt: graphics need to be sewn or woven into the larger pattern of meanings. The reader should be told their significance, why they are there and what they show, and this needs to be done at certain points in the narrative, and in certain ways. How and where you refer to your visuals can make a subtle but very important impact on the reader's experience of your document.

There are two basic ways to refer to figures in the text. One is indirect, with the reference placed in parentheses, for example, (fig. 1.2); this is the most common form. The other approach is direct and makes the graphic a subject of discussion, for example, "Figure 1.2 shows . . ." Deciding which form is best often means choosing whether you want your reader to glance over a particular figure, or study it in extensive detail.

No hard-and-fast rules exist to guide you in your decision. But some commonsense precepts can be applied. If a particular graphic is used to demonstrate or establish an important finding or to suggest a central interpretation or conclusion, it is appropriate (though not necessary) to mention it directly, making it the subject of one or more sentences, even a paragraph. On the other hand, if you include a visual to illustrate a point made in the text, to give an example, or to sketch a piece of apparatus, then indirect reference is probably sufficient.

Among our examples, a few figures that might deserve direct discussion are 9.2, 9.11, and 9.12. Each shows information that is central to the narrative, representing a major result of the experiments performed. Indirect reference, meanwhile, seems more appropriate to graphics like 9.3, 9.7, and 9.10, which provide generalized data or schematic representations. In the end, such decisions can often be subjective. If you have doubts about how to handle a particular graphic, look at your models or check the recent literature to see how other authors have dealt with similar images.

Having made your choice as to direct or indirect reference, where should you place it? In what part of a sentence or paragraph? Too many authors, when it comes to indirect mention (in parentheses), jump the gun and insert the reference before the reader is really ready. Examples abound in the literature. Here's one:

> The conventional structure of a bipolar transistor (Fig. 2A) requires three distinct material types: for example, a highly doped n-type layer . . . , a p-type base, and an n-type collector. (Cui and Lieber 2001, 852)

Note that the eye is told to abandon the sentence even before enough information is given for the figure to make sense. How are the authors using the figure? As written, they are emphasizing "conventional structure." But the actual message of the sentence and paragraph, and the figure too, focuses on the three parts of this structure, not its conventionality. Thus, it would be much better to place the figure reference either after the word "types" or, best of all, at the end of the sentence.

Another example relates to figure 9.7:

> Proved [natural gas] reserves estimated by the U.S. Energy Information Administration have risen from 1.4 BCFG in 1988 to 10.5 BCFG in 1995 (Fig. 2 [our figure 9.7]), more than 70% of which occurs in Colorado and New Mexico. (Murray and Schwochow 1997, 32)

Again, the reference belongs at the end of the sentence, after the pertinent information has been given. This cues the reader that *all* of the values mentioned are displayed on the figure, rather than reserves figures only. Note, too, that the reserve values given in the text are in BCFG (billion cubic feet of gas), rather than in thousand MCFG (million cubic feet of gas), as shown in the figure. This sort of discrepancy should be avoided. Using the same units in text and illustrations is needed to integrate the two.

Similarly, try to use phrases and terms in your captions that are picked up in the text. Much could be said (and has been) about how to write proper captions, what their length should be, whether to use full sentences or phrases, and so on. Standards here, as in so many other aspects, tend to change between fields and between journals. Make sure you're aware of any restrictions or rules given for your target journal or publisher. As always, where there is any doubt, use your models as guides.

A FINAL POINT

Scientific illustration, as it exists today in wondrous plenitude, can never be covered adequately by a single chapter such as this. Indeed, an abundance of volumes have been written on the topic, even on such seemingly humble forms as the chart or graph.[9] I've tried to cover some of the more obvious and necessary aspects of creating good graphics, and of evaluating those of others.

Beginning authors, once familiar with the literature, may feel (and even be thankful) that the types of illustrations available to them for their own writing appear to comprise a set of fixed formulae. Closer inspection will show that this is rarely, if ever, true. For many early-career scientists, it is helpful to stick to the most common patterns. But experienced scientist-authors understand that illustrations are actually

9. A few sturdy works in this field include Wolff and Yeager 1993, Anholt 1994, Briscoe 1996, and, of course, the several well-known books by Tufte: *The Visual Display of Quantitative Information* (1983), *Envisioning Information* (1990), and *Visual Explanations* (1997).

a flexible means of expression, and can be adapted, albeit conservatively, to individual cases. Close study of premier journals will prove this: different authors add modifications, sometimes small, sometimes significant, to what are otherwise standard graphical templates, and they do this in order to make their message a bit more efficient and elegant. Here, too, just as with writing itself, one can ultimately choose between functional and creative approaches.

CHAPTER TEN
Technical Reports

ORIGINS

Before the later part of the 19th century, the book held a central and hallowed place in scientific publication. Galileo, Newton, Lavoisier, Darwin—all presented the majority of their work in volumes of length and complexity. The book offered a scientific author the opportunity to cover many aspects of his observations, reflections, experiments, and theories in extended fashion, with considerable narrative sophistication and variety. The triumph of the journal has not extinguished this venerable forum of scientific writing, at least not entirely. But it has curtailed it significantly enough to have helped create a type of hybrid publication, a transitional form between book and article. This we know today as the technical report.

In the beginning, the report took shape as the monograph, a scholarly account mainly reserved for theses and special research studies. With the growth of industrial- and government-sponsored science, however, particularly in the 20th century and particularly in the engineering sciences, a different sort of venue had to evolve. Less academic in scope, more specialized in tone, able to embrace great quantities of detail, the technical report was born. Part reference work, part data reservoir, it has since grown to become a truly major outlet for scientific work. Today, reports of variable length are regularly contracted by a huge variety of entities—government agencies (federal, state, and metropolitan), companies (all sizes), consortia of one sort or another, research institutes, hospitals, the courts, neighborhoods, even individuals. Moreover, because of the tendency to outsource analysis and research these days, particularly in industry, the number and diversity of technical reports have gone through a burst of expansion.

Much of this work is kept out of the periodical literature, at least for some time. Indeed, large portions of contemporary science remain cooling between the shade of report covers, on shelves and in offices. This is one of the inevitable side effects of political realities in science, that is, "proprietary research"—the polite term given to contract and other work intended to be kept secret, whether because of industrial competition or governmental policies. In a fair number of fields—chemistry, engineering, petroleum geology, biotechnology—there is probably at any one time as much or more science held in this restricted domain as in the whole of the periodical literature. For researchers, this can mean a constant struggle, whether of heart or hand. The desire for open publication is countered by an administrative policy demanding that new knowledge be kept under house arrest. The broad truth is that proprietary research is an inevitable outgrowth of social and economic realities and has added enormous gains to human knowledge.

The story often has a happy ending (for science, that is). A great deal of confidential work does eventually gain release for publication. This can be very important for a particular field. Corporate labs and research centers, in particular, often have superior funding, top-of-the-line facilities, and high-quality research talent: the science produced within their walls can be advanced well beyond anything that academia can generate (one need only think of such entities as Bell Labs, IBM, or the major chemical companies). In my own field, geology, a great deal of frontline proprietary research has been performed by the major oil companies. At times, the release of this work has led to the creation of entire new subfields. In nearly all cases, such work first appears as formal company reports. From these, separate research papers can emerge, as well as oral presentations, abstracts, short course manuals, and even review articles (if a significant survey of the relevant literature has been done).

All of this highlights the importance of the technical report as a long-term container of valuable science. Whether the work involved is path-breaking or not doesn't really matter as far as potential publication goes. Reports can yield a harvest of publishing opportunities. Good reports are therefore written with this potential future somewhere in mind.

BASIC STRUCTURE AND FORMAT

Scientific reports vary in size from a dozen pages to multiple volumes, each containing reams of discussion, figures, data tables, appendixes, even a CD-ROM or two. They are thus a very flexible type of instrument, apt to be tailored to each specific project or to the demands of the contracting party. For some reports, a fixed and given structure must be followed; lab analyses and government reports are sometimes of this kind. More often, format details will be left up to the writers and compilers. Most formal reports employ a structure like the following:

Front matter
- Title page
- Table of contents
- List of figures
- List of tables
- List of abbreviations and symbols
- Abstract
- Executive summary
- Acknowledgments

Introduction and overview
- Introduction
- Project goals/objectives
- Purpose/scope
- Literature review
- Theoretical discussion

Main discussion
- Experimental methods, materials, and procedure
- Results
- Analysis and interpretation
- Summary and conclusion
- Recommendations/need for future work

End matter
- References
- Appendixes

This outline is a general guide, not a template. Different types of reports—feasibility studies, database reviews, research investigations, site visitations, project summaries—will require different shapes, some slim, some generous. The treatment granted an analysis of building-material strength may not be appropriate to fieldwork evaluating a new petroleum reservoir. In some cases, a literature review or theoretical discussion is not needed; in others, experimental methods or mathematical modeling are central. Reports that focus on the collection and analysis of samples (e.g. soils, materials, species) may require an individual section on collection methods. Contracted reports written for management or for clients often need a separate recommendations section. These are just a few of the many variations that pervade the vast universe of report writing.

It is usually quite helpful, and sometimes necessary, to have an example of a well-done prior report on hand when you organize and write your own. Using models is a constant theme of this book, and a very practical one. Whether you should use such an example as a general reference or a strict blueprint will depend on the type of report you're writing and who you're writing it for. In all cases, pay attention to any guidelines provided by the contracting party as you would the instructions to authors for a journal article (or any other set of biblical commandments). If no specific guidelines are given, keep the overall report structure flexible and open to change as you go along. Writers or compilers might decide beforehand, or at least consider, the general range and organization of information to be included.

But report writing is often just as experimental as the composition of a scientific paper (see chapter 4). Trial and error, dead ends, breakthroughs, and revelations should all be expected here, too. Good planning and organization will always help, no doubt. Yet these efforts are skeletal; they cannot do the actual writing for you or your team.

For longer reports, the main body should be divided into sections that represent individual data areas or topics. Each of these may be composed of work performed and written by separate project members or teams. It is then the compiler's job to make sure the order of these sections (as well as their literary quality) is not haphazard or "fractal," but creates a clear organizational flow. The compiler is therefore the one who should write the executive summary, which defines a very important window for the report. This summary should establish who the ideal reader is and what he or she should expect.

AUDIENCE: A CRUCIAL CONSIDERATION

Perhaps the most diverse and varied aspect of a formal report is its possible audience. This is one area where reports and journal articles definitely part company. Parts of your extended discussion of bacterial populations in nearshore mollusks might well be read by marine biologists, sewage treatment engineers, local naturalists, government officials, oil company executives, invertebrate paleontologists, epidemiologists, or the amateur clam digger. Each of these potential users will be interested in different sections of your report. A biologist will likely look at most of the main sections; a government official may use only the executive summary and conclusions; a medical specialist might focus on species identification and isolation techniques; the clam digger needs to know how abundant red tide or other toxic bacteria might be.

Is it possible to satisfy so many different readers? The answer, often enough, is yes—because each user comes to your report with a specific type of interest in what you have said. This means two basic things: first, your title, abstract, and executive summary need to be sufficiently accurate so that only those with such an interest will proceed. The worst sins of report authorship are discouraging appropriate readers and encouraging inappropriate ones; in other words, wasting people's time. Second, each section of your report must be sufficiently self-contained so that readers with specific needs can get what they need efficiently, and therefore appreciatively. Different users will have different information requirements and different amounts of time to spend with your work. A goodly number—some of them perhaps important to your future career opportunities—will read only one or two sections. You can't afford to make any part of a report weak or flabby. Such is tantamount to insulting a reader's interest and need—another cardinal sin. To put this more positively, it makes sense to consider the special power and advantage of a well-done report: it can be many things to many users.

REPORT WRITING AS A FORM OF TEACHING

This special power of reports partly resides in a unique opportunity for scientists to merge teaching and research. As I have said, acts of professional writing in science are instructional at their heart; they try to impart new knowledge. Reports do this at several levels. They provide a medium to educate readers as to the significance, theory, and existing literature on the chosen subject, and to provide the results of new investigation and their implications for past and future work.

The teaching aspect is especially relevant here, particularly to the introduction and overview section. This is where you must bring the reader up to speed, making her or him a capable and appreciative reader of everything that follows. Recall that there is no substitute for knowledge of the relevant literature: a good review here, preferably written early on, can provide invaluable focus and insight for your own work and also serve as a useful tool for others.

Consider, too, that reports of the past have a way of haunting the present; no topic is ever exhausted or consumed by a single project. Good scientists will always make a thorough search of what has been done before beginning their own work: otherwise, they are fated to repeat history in a very literal fashion. Moreover, there may be a political-economic dimension to consider: your work on solar collectors back in the 1970s may well prove useful once fuel shortages set in, prices rise, and the government again focuses significant grant money on alternative energy sources (this will happen, by the way).

Thus, try to think of your own report as both a record and a legacy. It is a document that should be legible to contemporary eyes and those 10, 20, and 30 years hence. This may seem a bit grandiose for routine lab analyses or one more site visitation—but it is not. Good work performed and recorded remains good work; it does not stop being "science" and therefore of potential use to others.

PRACTICAL MATTERS

The following points of advice are likely to help you at some stage of your report writing.

A major report must be planned, just like the research of which it is a part. If various teams are involved, it can be a big help to plot out the subjects to be covered and to consider the types of writing that will be involved. Will you need to include a detailed experimental section? A description of equipment, proteins, or rock types? A section or appendix on mathematical derivations? What types of illustrations might be needed? Who (if not yourself) will write or produce these parts of the report? You may not be able to answer all such questions at the outset, but it is important to put them in mind and re-pose them as you go along. The creation of a report needs to be managed, whether it is done by you alone or by a series of research teams.

If relevant, use any previous writings of your own (or your participants) as a source for composing individual sections. Above all, the proposal for your particular project can act as a literary reservoir. Very often, the introductory section of a proposal can be adapted (or even used directly) in the final report. If the proposal was especially long and detailed, several sections might draw from it.

Use of detail in a report is often most effective when both text and figures move from the general to the specific. This will be true both for individual sections and for the report as a whole (look again at the structural format outlined above). Don't overlook illustrations in this regard, as these typically compose a large portion of reports: in many instances, graphics carry the main data presented and thus need to be carefully designed and ordered to guide a reader visually through the material.

Think of your reader, who is coming to your work for the first time. Make his or her job as easy as you can. Provide a list of all abbreviations and define each one the first time it appears in the text—this is *not* a minor point, but can affect the legibility of your entire document. Use subheads fairly often to create pauses; a reader often needs to stop and take in what has just been said, to connect it with previous sections. Wherever possible, break up long sections of text with a figure or two. Seriously consider using an illustration to summarize—that is, reduce or eliminate—any portion of descriptive text that might lend itself to such substitution.

Keep in mind that, if written by several individuals or teams, a report will likely require an additional level of revision compared with a scientific paper. As the final compiler/editor, that is, you may well inherit a compositional Frankenstein monster. Your job will therefore be a big one, and of enormous importance: in your hands rests the responsibility for adjusting individual sections, rearranging them, and stitching together anew the total document. If performed seriously, this task can bring to life what essentially may have been lifeless literary tissue. On the other hand, if done hurriedly and without care, the very opposite can result.

It is a very good idea, these days, to put both text and graphics into digital form. This is a decided bow toward both the present and the future. While it is more than likely that text will be originally produced this way, such may or may not be the case with tables, charts, graphs, and illustrations. These, however, may be easily scanned into any number of digital formats. Putting your document in electronic form will certainly give you the greatest degree of flexibility in terms of revising, editing, and assembling it; sharing and comparing sections between different authors; and maximizing delivery options. It will also make your report potentially far more useful to others.

One of the report's greatest enemies is redundancy. This is especially true if the work of several researchers or research teams is included, each of which may have written their own part in the overall drama. A compiler or editor should eliminate as much obvious repetition between these sections as possible—keeping in mind, however, that a bit of rehearsal may be necessary to introduce each section and relate it to the overall scope and purpose.

A related problem is use of the report as a dumping ground. As with the journal article, more is not necessarily better in this context. Just because a report is larger in size and broad in scope does not mean that it presents an unmitigated opportunity to discuss every procedural step or to put forward every piece of data generated. A report needs to present a selection of work performed—a larger selection, to be sure, than in a scientific paper, but a selection nonetheless. True, some reports do require the inclusion of most data generated during a project. But that is what appendixes are for—to present information that is deemed important but that is not discussed directly in the text itself.

Beware the ominous deadline—every report has one, and you must plan for it. Keep any schedules up-to-date; meetings between research teams should be held regularly to discuss writing progress, areas of possible overlap, and any other important topics. A compiler or project manager should make sure that every member who will contribute to the writing gets a copy of any stylistic guidelines and report specifications. Everyone involved must be clear on what's needed and expected. No one should have to start from ground zero when it is time to do the actual literary work.

CHAPTER ELEVEN

The Proposal

It's hard to make predictions, especially about the future.
—Yogi Berra

THE IMPORTANCE OF PROPOSALS

Contemporary researchers tend to be worldly individuals in at least one respect: they know that a proposal does not always lead to marriage. Indeed, in the often financially polygamous world of modern science, it is necessary for a researcher to seek many unions, some temporary, some long-term, all based upon proper and well-expressed overtures.

The proposal is one of the more important technical documents written by scientists and engineers today. There are several reasons for this. The first, of course, is financial: proposals form the basis for evaluations that lead to funding (or not), and thus make a great deal of science possible. Second, writing a proposal forces you or your research team to step back and create a systematic plan for your work—to conceive and organize its activities, apportion responsibility, consider monetary needs and constraints, think about scheduling. Third, proposals very often serve as literary reservoirs: if done well, they can provide much material for future articles, for communication with peers, for talks and poster sessions, press releases, and so forth. Often enough, a proposal is the first real publication to emerge from a particular line of research.

MAKING A CASE

What, then, *is* a proposal? How should we think of it, as a form of writing? Some authors have called it—and many researchers would probably agree, in feeling—a sales document. But this is not quite right. Writing that sells is not promissory in a true, contractual sense; it aims at getting around careful evaluation through rhetorical superlatives that focus on one-time exchange—bait to the unwary. If we think of proposals as somehow linked to the idea of marriage, the sales document becomes a literary vestige of "the second oldest profession."

In practical terms, then, a proposal is usually something more interesting, and complex. It is several things: a *request* (for interest and funding), an *argument* (for the significance of certain ideas), a *blueprint* (for work to be performed), and a *promise* (that the work will be done within specified limits). Let us take these in order.

As a request, the proposal is commonly written to be evaluated by two or more researchers in the same field. This means that other scientists, not an anonymous funding agency, are your audience. Your document needs to have sufficient technical information and to be written in an appropriate style. These scientists' time is valuable (especially to them); they are getting paid nothing or very little to do this gatekeeper work, which they take very seriously, so be as direct, economical, and to-the-point as you can. Anything you say wastefully can and probably will be used against you. In short, your proposal should have dignity—it must embody the professionalism of the evaluators and acknowledge the pressures they face.

As an argument, meanwhile, your proposal identifies and outlines a specific problem. In most cases, this means a gap in the existing knowledge that (urgently) needs to be filled. The importance of this problem, what else depends upon its solution, is crucial to your discussion. You must convince your reviewers that your work will not only be worthwhile, but significant—that it will enhance the field (of which they are members) in some specific way. Nearly all the details presented in your document, then, should flow toward supporting this claim of significance.

As a blueprint, your proposal explains the work you are going to do to solve the identified problem. In simple terms, you have to show that you can create and present a reasonable, concrete research plan, that is, produce plausible time lines, budget analyses, equipment requirements, and so on. This often means, of course, that you are the maker of beautiful fictions (even fantasy), since it is always impossible (thank heaven) to constrain future research work in advance. Be assured: your reviewers know this. Most often, they will have generated proposals themselves and thus have dipped into the wax of "creative writing."

Finally, as a promise, your document is a pledge that you (or your team), personally, are serious, competent, and responsible, that you have thought things through and have the skill to carry out all work. This may sound self-evident, but the truth is that many proposals fail because they appear hurried or thrown together, don't follow directions provided by the funding agency (e.g. in the request for proposal), or are so poorly written as to give the impression that the researcher or research team just couldn't be bothered to take the time and care needed to produce a good document. This type of failure carries a promise of its own, of course (let us not speak its name), as well as a touch of insult. Yes, true enough, the hurried nature of many proposals reflects the conditions and means of their production—the dash to make deadlines, to invent details, to coordinate pieces by different groups, and to sew the whole together in Dr. Frankenstein fashion so that the result walks and talks without too many seams showing and without becoming a danger to its makers. Your proposal must indeed hide all this reality. But this is because the task is to guarantee that you will enter into a contractual relationship and abide by all the responsibilities involved.

In the end, therefore, a good proposal will make a case for both you and for your work. As put by Paradis and Zimmerman (1997, 116), it will reveal "the value of your idea, the elegance and good sense of your work plan, the strength of your preparation, the appropriateness of your facilities, and the economy of your budget." Another point, too, is often overlooked. The proposal shows that you or your team can write. It reveals that, if and when the occasion arises, you will be able to generate good-quality articles or reports for the benefit of other scientists and therefore the progress of your field. A poorly written proposal on an exciting topic is like a beautiful face with one eye.

DEFINITIONS AND REALITIES

Proposals come in two basic types, solicited and unsolicited. You write a *solicited* proposal when you respond to an announcement that money is available from a particular source, such as a government agency, corporate sponsor, or foundation, and is to be performed in a specific area or on a particular topic. Such an announcement is known as the request for proposal (RFP) or, in the case of engineering contracts, an invitation for bids (IFB). The RFP or IFB will nearly always provide guidelines, sometimes rather detailed, on how the proposal should be prepared. You write an *unsolicited* proposal, on the other hand, for potential sponsors that have not made any such formal announcement, but who have funding programs in place. In this case, the sponsor may or may not provide guidelines on how it wants a proposal to be written. Overall, solicited proposals enter you into direct competition with other scientists; unsolicited proposals do so only in an indirect way.

Researchers also need to understand the fundamental difference between grants and contracts. Basically, a *grant* is a form of financial assistance for approved work. A *contract* is a legal instrument by which a sponsor acquires (buys) certain services and resulting products. Grants, in theory, are like scholarships; contracts are much closer to purchase agreements. These differences may seem clear enough on the surface, but in fact they've always been a bit troubled, due in part to questions of ownership (e.g. when patents result from the research at issue) and how far such ownership extends. In recent decades, moreover, as the lines between "basic" and "applied" science have all but disintegrated in many fields, distinctions have become even more problematic. There are now certain crossover forms of sponsorship. One of these is the *cooperative agreement,* defined as a grant for work in which the sponsor (usually the government) will have significant programmatic involvement. Then, too, there are various *cost-sharing agreements,* in which a particular sponsor will put up only a specified portion of the total budget, with the remainder supplied either by the research institution alone or by a combination of other sponsors.

Obviously, given all this complexity, it is absolutely essential for you, the individual scientist, to become as familiar as you can with the types of research support common to your own field. Avenues open to the physical chemist are quite likely to differ from those available to the paleontologist or planetary astronomer. The institutional realities of professional science have become more heterogeneous—and more field-specific—with each passing decade, and there is every reason to think that this will continue in the future.

At the same time, however, there is one major trend that cuts across most fields. The past 40 years have seen solicited proposals come to dominate most scientific research globally and in the United States. This is largely due to the expanded role of government funding in science. Corporate sponsorship, meanwhile, has also grown, but is concentrated in certain fields, for example, biotechnology, computer science, energy-related research. In some countries, such as Japan, corporations account for nearly as much funding as the government, with both solicited and unsolicited proposals accepted. But in general, solicited science now rules over most of the industrialized world. This means that many nations have adopted or adapted the RFP system. Agencies, corporations, and foundations publish regular releases (annually or even monthly) announcing and outlining their research support. It is essential to gain access to these announcements, through hard copy and Internet subscription, if possible.

The rise in solicited science has brought with it certain effects. First is the fact of increased competition for available funds. This means that reviewers are now almost routinely overwhelmed with the number of proposals they have to deal with, and are thus pressured to be skeptical, impatient, and decisive. The strong lesson here is that your proposal should be an easy read—clear, to-the-point, aimed at giving the reader maximum content in minimum space.

Second, the importance of government funding has brought an increased dependence on political trends (commonly euphemized as "national priorities")—evident, for example, in the United States, with such developments as the "war on cancer" in the 1970s, "star wars" research in the 1980s, and, in more recent times, the defeat of the superconducting supercollider and the rise of the Human Genome Project. An ability to be nimble in taking advantage of public, governmental, or commercial priorities (including hot topics, such as specific illnesses) and avoiding others that have collapsed, can be an enormous asset in acquiring research funding. The able proposal writer will keep aware of such trends in his or her field, for it often pays to attach one's own research—without too obvious a stretch—to them. Knowing which way the wind is blowing is a definite means of survival, as any meteorologist or zoologist will tell you.

Third, government priorities also involve social considerations. In the last two decades, for example, many funding agencies have attempted to inaugurate grants specifically available to minority researchers and related institutions. There are also now grants for research that promises to have some positive effect on public understanding of science or on teaching and training.

Another important reality is the increasing role of the Internet. As of this writing, the vast majority of grant-giving agencies have Web sites where they provide all essential information and, often enough, application forms to potential grantees. In view of this reality, certain groups have collected such sites, and the updated information they provide, into centralized databases and have made them available to scientists, usually on a subscription basis (two such enterprises, each with over 200,000 members worldwide and tens of thousands of grant opportunities, are the Community of Science, Inc., found at www.cos.com, and Sciencewise, Inc., at content.sciencewise.com). Other companies have provided software for purchase to search a large single database of funding sources (an example here, specifically for U.S. federal grants, is Federal Money Retriever, available at www.fedmoney.com). These are moves to take productive hold of the future. Any such enterprise, if well handled, can provide an invaluable service.

Individual agencies, meanwhile, are now accepting—and in some cases, like the National Science Foundation (NSF), even *requiring*—that proposals be submitted electronically. This has clear benefits in terms of efficiency, cost, ease of handling, tracking, and archiving. Because of this, online submission will very likely become the standard within a decade or less. If at all possible, therefore, you should try to submit in this fashion, as it is clearly the wave of the future. The advent of e-science is very much upon us, and proposal writing is an essential part of this.

MERIT CRITERIA

Scientists who review proposals are like editors. They must choose between competing entries, and they must do so under pressure. They are guardians of quality who are very much aware of this role and are deadly serious about it. Part of this seriousness involves the criteria they commonly employ to judge merit. Sponsors often provide reviewers with such criteria, and, as responsible professionals, reviewers will certainly use them—but not without other considerations.

If you were to sit down (as I have, for the purpose of this book) and ask reviewers from a range of different fields just how they evaluate an individual proposal, what they look for to award high marks, you would find a list like the following:

1. Scientific content: How good is the science involved? What questions or problems does the proposed research set out to solve and how knowledgeably are they presented? (Proposers must show adequate command of the relevant material.)
2. Significance/importance: Why is this research needed? How will it advance knowledge in its particular field? To what degree is it truly

original or does it fill a specific gap? Where might it lead in the future? (The problem or question must be interesting, therefore defensible on grounds of significance.)

3. Feasibility: Is the proposed research practical? Can it be performed such that the specific question or problem can be answered within the given time period? (Many scientists can pose interesting questions, but some problems will inevitably be years away from being answerable.)

4. Clarity: How well does the proposal read? How simple and straightforward is it? How often is it necessary to stop and try to figure out what is being said or meant? (The proposers must be able to express their ideas and information adequately to others.)

5. Flow: How well is the proposal organized? How logically do the various parts fit together? Is there unneeded repetition? Are there gaps that force the reviewer to worker harder than necessary to understand the proposed work or thought process? (The proposal should help the reviewer do his or her job easily, with a minimum of extra effort.)

The first two criteria will be found at the top of every sponsor's list of evaluation guidelines. Quality and importance of the science are primary: this is a given, across the board. The question of feasibility, meanwhile, is something that an experienced reviewer will always look upon as crucial, whether this criterion is specifically given by the sponsor or not (often it isn't). If the project isn't realistic, it probably shouldn't be funded: this, too, is a given. Thus, reviewers are concerned, first of all, with technical questions. But they are also influenced, to no small degree, by the type of reading experience a proposal creates, and this is where the last two criteria come to bear.

In recent years, some granting agencies have incorporated other merit criteria into their programs. The NSF in the United States, for example, and certain public support agencies in Britain and Europe, now ask that projects be justified in part on the basis of some of the following criteria:

- How well could the project help integrate research and teaching?
- Will results be made available to a broad audience to help advance scientific and technological understanding among the public?
- Does the research involved enhance in any way the participation of minority groups?
- Are there any clear benefits of the proposed research to society at large?
- How interdisciplinary is the project, and does it help advance this type of research?

- In the case of grants given to scientists in developing nations, how well does the project strengthen or advance "endogenous" basic science, that is, provide incentives to reduce the exodus of research talent to developed nations?

One or more of these criteria might apply to any particular proposal. As always, part of your responsibility as an applicant is to know what a specific organization is looking for. Some idea, beyond the details given above and those found in any agency proposal, can be gleaned from a visit to the organization's Web site.

In all cases, however, a good proposal is written not just to "get things down on paper," but to create real interest. Moreover, it should not only create such interest, but leave it unsatisfied, in anticipation. If a scientific article tells a story, then a proposal does too, but without an ending. It will make the reader want to see the work done (and done well), to see how it all comes out. And this is where the quality of the writing itself comes in.

Reviewers see your proposal as an indication of how good a researcher you (or your team) are. If your presentation is cogent and logical, if it doesn't waste a reader's time with extraneous details or confused wording, you will appear in control of your subject. Furthermore, how well you respond to the specifications given in an RFP also tells a reviewer or sponsor something important: whether you can follow directions—and thus carry out the proposed work.

Thus, anything that might make your document clearer, more to-the-point, better organized, and precise is worth considering. As in all other forms of scientific writing, the process of revision is where the ore is refined into gleaming metal. Ask yourself, at every stage or step, whether a particular point or detail is really needed, what it adds to the total, and whether it contributes toward the argument and toward the appearance that the work is important, well-planned, properly budgeted, and thus in the right hands.

Because it is a request for money, a proposal should project a sense of confidence, realism, and solid planning. Reviewers should be led to feel that any funding will go directly toward efficient completion of the project and related publications.

USE OF MODELS (AGAIN)

Let us return to one of the main themes of this book. With proposals, too, good models are invaluable, no less than in any other area of scientific writing (and perhaps even more, given what's at stake). One reason why models are especially important here is that scientific proposals come in an enormous variety of shapes and sizes, matching in a sense the varied institutional dependencies of science itself. In a number of fields, including my own (geology), they range from brief, ten-page documents to research bids that stretch to several volumes and thousands of pages. Becoming intimate with proposal formats in your own area qualifies as a necessary form of literacy.

Choose examples to study and imitate that were not only funded but that received high marks from reviewers. Veteran colleagues are one of your best resources here. Ask them to supply you with samples they consider particularly worthy—then ask them why they consider them so. Such samples might include proposals these colleagues wrote or worked on themselves or others they reviewed (many scientists keep copies of proposals they've given high marks to). Use common sense in choosing your models. Be sure that they are recent—proposal requirements change over time—and that they targeted sponsors you yourself might select.

Also think about collecting one or two examples of proposals that were poorly done—where the problems were clearly identified and commented upon by reviewers. Learning what to avoid, or at least watch out for, can be as useful as finding what to emulate. Or, in more blunt terms: nothing succeeds like avoiding the errors of others.

In studying your positive examples, pay attention both to big questions and to details. Big questions include, How much knowledge of the specific topic was assumed? How are the main goals laid out? How is the reader urged to see this as a truly significant and original project? More detailed points to consider might be, How long is each section? How is it organized? What type of style was used; for example, did the author employ questions, subheads, enumeration, or other aids to guide the reader? What specific parts of the proposal seemed to you especially strong, well-written, concise?

Besides studying good and bad examples, you can learn a great deal from experienced colleagues through conversation. As noted, colleagues who are experienced proposal writers, and who have served as reviewers, are a great resource. Sit down and talk with them about their perceptions. Ask them what impresses them about a proposal, what common blunders they see, what makes them thankful or frustrated, what points they would stress were they to give a lecture on the subject. Chatting about the whole business can bring up practical tips that might otherwise not be evident from the written example alone. Most important of all, use your colleagues as first-pass reviewers: have them read through your own proposals and make suggestions.

Finally, keep your eyes and mind open. Whatever models you choose to start with, consider adding new ones at a later time if you come across any that seem better or more appropriate. Writing, as I've said many times in this book, is an experimental process, which means a process of continual learning. As your proposal-writing skills advance, your needs for improving and refining these skills will naturally change. Good writers are always interested in adding new tools to their workbench.

EXAMPLE

The succeeding pages of this chapter offer portions of a well-written proposal that gained high marks from reviewers and was therefore funded (by the National Institutes of Health, a major research agency in the United States). It concerns research in the field of immunology and is highly technical. The alert (or even not so alert) reader will note that I have retained a significant amount of the original vocabulary—this is done for specific reasons. First, most scientific proposals *are* highly technical, indeed they *have* to be, thus any other type of example here would be superfluous. Second, it is very important to be able to "read through" this terminology in order to understand the actual patterns of language flow and the logic involved. Only parts of several sections are included (the original was over 30 pages long).

To begin, let's look at the description section (fig. 11.1). A majority of proposals today ask for this type of abstract at the beginning and may specify what it should include: long-term objects, specific goals, basic research design and methodology. As shown, the abstract must fit into a given space, which demands that it be no more than about 400–500 words, and often less.

Note in the example that the very first sentence presents a short, concise statement of the overall goal. There is a sensitivity to the reader here—a simple, straightforward entry into the project. This is followed up by an enumerated listing of specific aims, surveying the What of the proposal. Next comes the How, a brief discussion of approach/methodology. Finally, we are told the Why—first in terms of specific significance ("discovery of genes involved in class I and class II restricted antigen processing"), and then related to more general importance ("defenses against microbial pathogens" "pathogenesis of autoimmune diseases"). The whole thus forms a neat closure, moving from What to How to Why, and at the same time starting with the general, moving

BB Principal Investigator/Program Director (*Last, first, middle*): ________________

DESCRIPTION. State the application's broad, long-term objectives and specific aims, making reference to the health relatedness of the project. Describe concisely the research design and methods for achieving these goals. Avoid summaries of past accomplishments and the use of the first person. This description is meant to serve as a succinct and accurate description of the proposed work when separated from the application. If the application is funded, this description, as is, will become public information. Therefore, do not include proprietary/confidential information. **DO NOT EXCEED THE SPACE PROVIDED.**

The overall goal of this grant is to identify new genes involved in antigen processing and presentation. The specific aims are: 1) to search for genes that affect the processing and presentation of HLA class II restricted antigens; 2) to search for these genes in the MHC, in unstudied flanking regions of chromosome 6p, and in the rest of the genome; 3) to identify new genes that affect the processing and presentation of class I restricted antigens; 4) to investigate whether the MCH-linked heat shock genes have a role in antigen processing and presentation; and 5) to investigate the basis for the alteration in recognition of antigen processing/presentation mutants as targets for alloreactive T cell clones. The basic approach to new gene identification will be to isolate mutants affected for processing/presentation, using selective schemes designed to distinguish between antigen processing-competent and -incompetent cells. . . It is anticipated that these studies will lead to the discovery of genes involved in class I and class II restricted antigens processing. In a more general way, this project should enhance our understanding of the mechanisms involved in antigen presentation to T cells, mechanisms that are critical in host defenses against microbial pathogens and that may play a role in the pathogenesis of autoimmune diseases.

PERFORMANCE SITE(S) (organization, city, state)

KEY PERSONNEL. See instructions on Page 11. *Use continuation pages as needed* to provide the required information in the format shown below.

Name	Organization	Role on Project

PHS 398 (Rev. 4/98) Page 2 BB

Number pages consecutively at the bottom throughout the application. Do *not* use suffixes such as 3a, 3b.

Figure 11.1 The abstract portion of a grant proposal

to the increasingly specific, and outward to the general again. This type of hourglass approach (broad, narrow, broad) is among the most effective rhetorical techniques for guiding readers through any project, and for giving them the sense that they have grasped the essence of the matter.

The budgetary details show a common type of breakdown for direct costs (fig. 11.2). Note that the requested salary amounts are calculated, essentially, on the basis of person-months: this is typical for many proposals in the United States and elsewhere. Almost all proposals will ask you to break out specific equipment and supply costs, as well as any outside consultant or service (e.g. computer processing) costs. It is also expected—and indeed, might be advised—that researchers include costs of travel to conferences where they will present the results of the funded research, and also page charges for publishing this work.

Now consider the research plan. Here is where the authors flesh out details regarding the aims, background, and significance of their project. In our example, the specific aims are listed, as they are in the description (fig. 11.3). The wording is not exactly the same in both—note that a bit more detailed information, including terminology, is given here, especially in the last two items. But the same order is kept, indicating that the authors are aware of their earlier pattern and presentation. Moreover, the degree of specificity tends to increase toward the bottom of the list—again, the pattern moves from more general to particular, which is an excellent way to guide your reader into the matter of your discussion.

This same pattern emerges again in the next section of the plan, on background and significance. Here is one of the most important parts of any proposal, since it presents an argument for the rationale, intelligence, and ultimate value of the project. Succeed here and your readers are your allies. Notice that, at the very beginning, the authors outline the gap in knowledge they will try to fill. They do this, moreover, by stating what is known and what is not known, first very generally (in the first few sentences), and then, in more detailed fashion (second paragraph). Questions are nicely used: not only do they draw the reader in, emotionally and intellectually, but in their particular order, these questions offer a chain of logic that suggests a well-reasoned approach to the research at hand.

Finally, in the third paragraph, the significance of these questions is brought back to the larger realm. The authors place their thinking in the context of ongoing work—they admit their debt to others and, at the same time, include themselves in the community of researchers on related topics—and state that what they are really after is new

Principal Investigator/Program Director (*Last, first, middle*): Newton, Isaac

DETAILED BUDGET FOR INITIAL BUDGET PERIOD DIRECT COSTS ONLY					FROM 9/01/2002	THROUGH 08/31/2003	
PERSONNEL (*Applicant organization only*)					DOLLAR AMOUNT REQUESTED		
NAME	ROLE ON PROJECT	TYPE OF APPT. (months)	% EFFORT ON PROJ.	INST. BASE SALARY	SALARY REQUESTED	FRINGE BENEFITS	TOTALS
Isaac Newton, LLD	Principal Investigator	12	24	125,500	31,375	7,450	38,825
Robert M. Hooke	Co-Invest.	12	50	96,500	48,250	8,200	56,450
John R. Keill	Sen. Fellow	12	75	65,500	49,125	5,100	54,225
Samuel Clarke	Sen. Fellow	12	50	60,000	30,000	3,750	33,750
Gottfried W. Leibnitz	Research Tech.	12	100	30,500	30,500	1,250	31,750
		Subtotals ⟶			189,250	25,750	215,000
CONSULTANT COSTS							
EQUIPMENT (*Itemize*) Centrifuge ...... 11,000 Spectrometer 8,000							19,000
SUPPLIES (*Itemize by category*) Serum .... 9,500 Tissue culture media .... 2,100 Chemicals, enzymes .... 5,800 X-ray film, filters .... 1,300 Glassware, plastic .... 4,100 Liquid nitrogen, gases 1,800			Synthetic oligonucleotides .... 1,500 Isotopes, labeled compounds .... 5,600 Small equipment items and misc. supplies 3,794				35,494
TRAVEL Two trips to national meetings by PI and one co-investigator							6,500
OTHER EXPENSES (*Itemize by category*) Page charges .... 2,500 Books, journals, software .... 2,000 Protein sequencing .... 1,000 Equipment maintenance .... 2,500 Use of Flow Cytometer .... 2,000 Telephone, postal 500							10,500
TOTAL DIRECT COSTS FOR INITIAL BUDGET PERIOD ⟶						$	286,494

Figure 11.2 The budget section from a grant proposal

knowledge that will advance "our understanding of immune system response," especially in the thymus gland.

All of this, together, offers an auspicious beginning. It is clear to a reviewer that the proposers can do several important things: they can think through their ideas clearly, creating a solid, rational plan for them, and they are skilled enough as authors to generate good-quality publica-

Principal Investigator/Program Director (*Last, first, middle*): ____________

RESEARCH PLAN

A. Specific Aims

1. To search for genes that affect the processing and presentation of class II restricted antigens, specifically in portions of the major histocompatibility complex (MHC) and in unstudied flanking regions of chromosome 6p.
2. To search for genes, affecting class II restricted antigen processing and presentation, which map elsewhere in the human genome than on the short arm of the 6th chromosome.
3. To search for new genes that affect processing and presentation of class I restricted antigens.
4. To evaluate what role, if any, the MHC-linked heat shock genes and their cognates may have in antigen processing and presentation.
5. To investigate the basis for the alteration in allorecognition of antigen processing/presentation mutants by alloreactive T cell clones.

B. Background and Significance

Although researchers have gained many important insights on the biology of antigen processing and presentation, little is yet known about the specific genes and gene products that carry out these functions. This is particularly the case for class II restricted antigens, which have a crucial role in immune system response. Prior research has confirmed, however, that the overall process is quite complex [references]. At a minimum, it involves several main aspects, including: the biosynthesis of class II α and β and invariant chains in the endoplasmic reticulum; the association of these chains in a nine-unit complex; and the trafficking of this complex through the Golgi to the trans Golgi reticulum, directed by an Invariant chain signal, and thence to the endocytic pathway, where the Invariant chain is degraded. . .

While this scheme is probably correct in general terms, many of the details remain unclear. For example: Where and how do peptides derived from lysosomal processing become associated with class II molecules? Do class II molecules recycle through the endosomal compartment and acquire newly generated peptides for presentation, and, if so, in what cells? In what compartment are endogenous membrane proteins degraded to yield the peptides which associate with class II molecules?

Work from several labs, including our own [references], suggests that the answers to these questions are very closely related and will have an important effect on our understanding of immune system response. In particular, the peptides that derive from endogenous proteins constitute a major class occupying the binding grooves of cell surface class II molecules, and appear involved in positive selection and the induction of self-tolerance in the thymus.

CONTINUATION PAGE: STAY WITHIN MARGINS INDICATED

PHS 398 (Rev. 4/98) Page ______

Number pages consecutively at the bottom throughout the application. Do *not* use suffixes such as 3a, 3b.

Figure 11.3 The research plan for a grant proposal

tions from the relevant work. Their claims are not too grandiose, nor are they overly modest. They project confidence, competence, and realism. Much of the rest of the proposal—including a section on research design and methods—takes up the task of demonstrating that the relevant work is feasible, and can be done within the budgetary limits presented. A reviewer is left with the feeling (however imaginary) that any funding will go directly toward efficient completion of the project and related publications.

FINALE

In the end, there will always be a fictional aspect to proposal writing. How could it be otherwise? Science is involved in many marriages, with institutions, social trends, political realities. Proposals are a way of admitting these dependencies by telling plausible stories about the future, always an uncertain reality. These are stories we tell to fellow travelers, other scientists. They are different from those related in articles, reports, and presentations; they have their own demands for believability—demands that are linked to the practical, competitive realities of science in the world today. To advance knowledge, in part, means convincing others that our work has been, and will be, valuable, well-conceived, and worthy. This, in itself, would seem a worthy skill to cultivate.

CHAPTER TWELVE

For Researchers with English as a Foreign Language

ENGLISH AS THE LANGUAGE OF SCIENCE: A FEW REALITIES

"Science and technology now have a true international language, and it is English." How true is this statement? In fact, it is very much true, but with certain limits.

Scientists must learn to read, speak, and write English to be fully active members of their profession in a global sense. This may seem a burden, especially to those from less wealthy nations where science must struggle to keep pace in many other ways, too. But history reveals, beyond any doubt, that this kind of situation has *always* existed in the past: people interested in scientific subjects had to learn Greek or Chinese in ancient times in order to have access to the most sophisticated thinking; in the early medieval period, it was Arabic (or Chinese again); then Latin from late medieval times to well after the Renaissance. During the 19th century, many researchers from America and elsewhere traveled to German universities and laboratories to gain knowledge of the most advanced methods and theories, especially in chemistry. The dominance of certain cultures and countries in the past have helped make them centers of scientific thought, to which others have had to journey.

If English is the main language of science today, this cannot be surprising. David Crystal in *The Cambridge Encyclopedia of the English Language* (1995; a fairly authoritative source on these matters) offers several reasons for learning English that are related to science. These include (1) intellectual reasons, since most of the world's technical and scholarly literature is now communicated in this language; (2) political-economic reasons, related to the dominant position of the United States in business, trade, the military, diplomacy, and other sectors; (3) practical reasons, because English is now used in publishing generally, in the computer industry, and in transportation; (4) entertainment reasons, because English has become the main language of popular culture (music, film, video, toys, etc.).

If you are a scientist or engineer, these reasons will all affect you and your professional life in some way. Scientific work today is written and spoken in English above all, and this has become even more common with expansion of the Internet, including the use of e-mail. Not only is the great majority of influential journal literature in English, but international conferences are routinely held in this language. In order to keep up with advances in your field, contribute articles and talks, and keep your professional opportunities open, you must learn English.

That being said, what English are we talking about? American? British? Indian? A combination? These all have significant differences, as any linguist will tell you. To be safe, however, scientists should focus on learning to read, speak, and write American or British English, as these will help guarantee the widest access to the written and spoken material of science.

READ, SPEAK, AND WRITE

Please note the order in which I list these skills—read, speak, and write. Reading is generally the easiest to learn. Speaking, at a functional level, is more challenging. But writing, for scientific publication especially, is hardest of all. Why? Not because writing is much more difficult or complicated, but because the standards for satisfactory work are very high.

To write a publishable paper, you must make your document as clear and as free of errors as possible. Wrong terms, missing words, or confused grammar will damage or obscure your meaning. Reviewers and editors cannot afford to be forgiving. Their job is to make sure that any material they accept is timely, significant, and, above all, readable. They cannot favor poor writing, even on an excellent subject, because doing so will hurt their own credibility and the professional status of the journal. Remember that, unlike speech, writing cannot be changed once it is in print: it is there forever and for all to see, again and again (and again). To write for publication is therefore more demanding, as a skill, than reading or speaking in professional situations.

LEARNING TO READ IN ORDER TO WRITE

How does one learn a foreign language? First, by imitation, and later, by improvisation. We learn to copy sounds, to memorize words and rules, to repeat phrases and patterns until we can easily create new combinations. We study examples of good expression, how their sentences go together, and we try to follow their example. In short, we learn to reproduce the language in its native forms until these forms become native to us (or at least partly so). It works the same way in writing science.

You begin by learning the words and grammar of English itself: this is a necessary starting point (I assume, if you are reading this, you have already come most of the way). Next, you learn the vocabulary of your field, the terms and technical phrases that are commonly used to discuss important phenomena. At the same time, you see how this vocabulary is used. You see it in articles and reports by scientists whose first language is English. You absorb the "dialect" of your field so that you can express your own work in it later on.

How do most scientists acquire this "dialect"? Mainly through reading. Over a period of years, constant reading of the literature, day in and day out, helps develop a sense of what sounds right and what doesn't. The process is by no means guaranteed; it tends to work for some people and not for others. It is very often slow and incomplete.

It can be done, however, much more quickly and efficiently. A starting point is to think of acquiring the "dialect" of your field as a part of learning English itself. Normally, of course, learning a foreign language involves memorizing grammar, reciting dialogues, doing exercises, taking tests, and so on. In the case of scientific language, however, you don't need to suffer through all this (again). But you will do well to follow a type of program that helps focus your attention in certain ways.

This program is based on finding and using examples—examples of the best-quality writing in your field. Here are some suggestions about how to begin:

Make a copy of excellent articles you come across and keep them in a file.

- Make sure these articles are simply written and easy for you to understand.
- Do not copy articles because they seem complex or because they are written by the best-known researchers, but because they are clear, well-organized, and understandable.
- Ask your colleagues or professors to recommend any articles they think are especially well-written and well-organized.
- Make sure these articles are recent, preferably less than five years old.
- You can also use chapters from recent books, including textbooks, as long as they are in your field.
- If possible, have a colleague who is a native English speaker look over your chosen articles/chapters and make any recommendations about which are the best.
- Keep the total number of copied articles or chapters fairly small (e.g. 15–25).
- Above all, try to find one or more articles on subjects that are as close as possible to your own work.

Reread these examples of good writing on a regular schedule.

- Take 30 minutes or so of each morning or evening to do this.
- Concentrate on one article per week (or longer).
- Take time, now and then, to listen to the words as you read them silently.
- If your memory is strong, try memorizing chosen paragraphs.
- Keep a list of terms you needed to look up: for each entry in this list, write out the sentence in which the word occurred.
- Review your list at the end of each week.
- Eventually choose one or two articles whose style and structure might act as a guide to help you write your own paper.

Imitate your examples of good writing.

- After your daily reading, take a few minutes to copy out paragraphs from your chosen article.
- Copy the same paragraphs for several days at least.
- Do this for an article whose subject is very similar to your work.
- Try to add a few sentences of your own (using fake information) to each paragraph or replace a sentence or two within the paragraph.
- Try to write a new paragraph in the same style as this article, using your own data (or fake data).

- Have someone else, preferably a native English speaker, read over these short writings and comment on them.
- Think of these activities as part of your language training.

These activities all have a single goal: to help you develop a sense of what sounds right in the language of your field and what does not. You may not have time to do all the things listed above. At the very least, however, it will be an enormous help to read and study good examples of scientific prose—examples that you have chosen yourself or that have been recommended by colleagues and that you find simple, clear, and worthy of imitation.

HELPFUL AND UNHELPFUL WORRIES ABOUT USAGE

It is a very common belief among scientists who are foreign speakers of English that the more they know about grammar, the better they will be able to write. In other words, the more rules they memorize, the more mistakes they will avoid and the better their usage of English will be.

Such beliefs are incorrect. Learning to write well doesn't happen this way. Knowing the rules of music notation will not make you a pianist or composer. Knowing all the rules of soccer will not make you an excellent player. The ability to write well comes from a different type of understanding, a kind of trained ear for good expression. This is true for writing in any native language; it is even more true if you are working in a second or third tongue.

Many books try to teach good scientific writing through rules and examples of proper usage. Most of these books are for native speakers of English; a few are not, but contain the same material in slightly simpler form.[10] Nearly all of them claim that the literature of science is badly written and desperately in need of emergency care. They imply, therefore, that it is a mistake to imitate the writing of other scientists. Such is an author's way of saying that you should rely on his or her set of rules for good usage. In other words, it is an advertisement. But it is something else as well. Stating that scientific writing is bad, in general, is a way of telling you not to rely on your colleagues in this area. It also suggests that nearly all editors are poor at their jobs. Neither of these points is true—nor could they be, unless science were about to disappear from the face of the earth.

10. An example of such a book is Yang 1995.

It can be a help to read through and study a book or two on scientific style and usage. Such books can help train your ear in a certain way, guiding you away from some common errors (I especially recommend Vernon Booth's *Communicating in Science,* second edition, because it is short, clear, and practical).

But this kind of book will never actually *teach* you how to write. This learning can come only from close, intimate contact with what others have successfully produced, especially in your own field. Good prose, not a rule book, will be your best teacher. You will find, moreover, that it is much easier to learn acceptable writing from the published literature than from a guide on scientific style.

THE VALUE OF COLLEAGUES

Science is very social in its work. Scientists interact with colleagues at almost every level and on a constant basis. Some of this interaction is formal, as in seminars, meetings, conferences, and so on. But most exchanges are daily and informal: when we have discussions in the lab or field, drop in at someone's office, chat in the hallway or on the phone, go out for lunch or a drink, exchange e-mail. A great deal of science gets done in these ways.

Colleagues are very often our most important resource. This is especially true when it comes to writing. You *must* have others read over your work if you are going to submit it to a journal or other publisher. These individuals, whether friends or fellow workers, will be your first editors, and you should think of them this way. If at all possible, get a native English speaker to look over what you've done and let you know how it reads, where it needs improving. Ask your readers to evaluate your grammar, style (readability), and organization (how well ordered are your sections and major points).

Remember, however, not to ask too much of anyone willing to read your paper. For example, don't request that an English-speaking colleague correct every error of grammar or spelling. Request that your readers spend only as much time on this task as they can safely spare—for if you are too demanding, they will not want to help you in the future. It might save time to show them several early sections of your paper first, rather than an entire draft. Their initial comments will help you improve your writing throughout the later parts of the paper, thus minimizing any changes you'll need to make there.

A BIT OF PRACTICAL ADVICE

Most of the earlier portions of this book contain advice on how to put together a scientific paper, report, or proposal. The chapters are simply written and should help the foreign speaker of English, too. Here, however, are a few essential points that might be emphasized for the sake of such a researcher.

If you are writing a paper for publication, you first need to choose a journal or journals where you wish to submit your work. Closely study several issues of your first choice. Pay attention to the organization of articles, their length, and their style. Copy the instructions to authors and refer to them as you write. Obey all style guidelines.

Be conscious of whether articles in your journal (and field) must all follow a given order of sections (e.g. introduction, materials and methods, results, discussion, references). Many fields, and therefore journals, do not have this type of standard.

If you find it difficult to start writing, find an article on a very similar subject and rewrite the opening paragraph, using your own specific topic and information. Remember that this is a way to help you begin; it is *not* a method for writing an entire article (using it this way may bring accusations of plagiarism). If you do copy an entire paragraph or two and replace only a few words, go back later on and rewrite it so that it is more your own.

Anytime you are uncertain of how to say something, or when you are stuck and can't seem to write anything more, take out one of your article examples (on a subject close to your own) and read it over several times. See if there are any phrases or sentences you can use to help get you past your sticking point.

THE NECESSITY OF PATIENCE

Learning scientific English is part of learning English generally and requires patience. It may take as much as a year following the type of program discussed above before you begin to feel comfortable writing the language of your field. Writing good scientific papers is no easy thing. Indeed, good scientific writing in English is quite foreign to many native speakers, too—that is the reason for a book like this one.

Please note that it has been shown, time and time again, that learning a language is best done when study is performed often, a little at a time, on a regular basis. Massive doses of studying and memorizing once a week (or even less often) are ineffective and inefficient. So is studying for hours every day. The same is likely to be true with regard to writing, which demands a particular ear for language that comes from long-term, thoughtful exposure.

Keep in mind, therefore, that improving your writing will require patience and time. Immersing yourself in books and articles will not make the process happen more quickly. On the contrary, it may lead to disappointment when you see that results do not come as rapidly as you hope or expect. Repeated study of single good examples is the best way to train your sense of what sounds correct in the writing of your field. Think of it as a form of training, where repetition is crucial.

CHAPTER THIRTEEN

Oral Presentations

A Few Words

THE SPOKEN WORD IN SCIENCE

Talking is one of the things that human beings do best. This is true, even of scientists. While no small difference separates a hallway chat from a formal talk, both involve sharing knowledge through the spoken word, the most basic form of communication. An oral presentation is an opportunity for you to be the only local speaker in science's unending conversation. One side of the conversation falls silent for a few moments, with interest and attention, to hear about you and your work.

It is not generally known that the beginnings of modern science largely derive from oral sources. The very first journals, which began publishing in the 17th century—the *Journal des Sçavans* in Paris and the *Transactions of the Royal Society of London*—were composed mainly of transcribed lectures. These lectures were given before each society, the Académie des Sciences in France and the Royal Society of London in England, usually at the rate of one per week and in a seminar format. Members were free to ask questions, offer criticisms, and discuss particular points. The original speech would then be revised and submitted for publication in an upcoming issue. Though each lecture was written out beforehand, it is likely that much ad-libbing went on during actual delivery. Writing and speaking were inseparable, like color and taste in an excellent wine. The tradition was carried on well into the 19th century, when the great rise of journal publication began.

Today, oral communication is integrated into scientific work at every level—think of classroom teaching, seminars, office banter, parties, departmental meetings, as well as conventions and symposia. We are all, in a sense, professional talkers in science. Consider: during a single week, you are likely to give any number of little "speeches" about your work, whether to a colleague, adviser, student, administrator, someone in another discipline, a family member, or the bathroom mirror.

Oral presentations take this very human reality and give it a more official, ceremonial cast. You may feel, sometimes, in a moment or two of stage fright, that such talks are the penance to be paid for a poor career choice. But this is only momentary anxiety speaking. Formal talks are the special opportunity we give ourselves to discuss our private work in a public voice, to make the labor of our minds and hearts communal.

ESSENTIAL ATTITUDES

Listening and reading are very different activities when it comes to absorbing knowledge. If written material presents a chosen residue, the spoken word is a communicational volatile: once offered, it vaporizes. Writing is static, permanent, and allows repeated study. Speaking is dynamic, transient, and reliant on human interaction. As a result, the "stories" we tell by voice need to be different than those we narrate in an article or report. And part of this difference resides in the fact that you, the speaker, are putting a human face on things, a face hopefully of enthusiasm and interest.

The professional talk is a kind of lecture. For a few moments, listeners are willingly transformed into learners, and you become their instructor. This does not mean they are available for condescension—they are your equals in every sense except with regard to the specific knowledge you are about to impart.

Is your presentation a type of performance? Yes, of course—but not merely that. You don't have to deliver a presidential address to do your work well. But it will help your message enormously if you give it *some* personality. Your audience is already captured, so to speak; lecturing with a degree of enthusiasm is being considerate and professional. Dull and monotonic recitations are a type of insult to those who are spending money and time (away from their own research, their families) to come listen. If your voice and manner display interest, your audience will, too.

Why do I emphasize these realities? Because they can help give you direction. If you are a teacher as well as a researcher, think of an oral presentation as a natural extension of your classroom efforts. If not, consider it an opportunity to practice teaching skills that might serve you excellently in the future (e.g. if you decide to make a career change).

Here are some points to consider:

- Begin to *prepare* your talk *well in advance*. Outline a basic structure for it (use headings), consider alternatives; think about visuals. Take notes on other talks you see that you consider especially good.
- Design and write out your talk in a manner you *feel comfortable* with, whether this involves using an outline, note cards, storyboarding (laying out visuals in order, with a bit of text for each to guide what you are to say). Experiment, if necessary, to see what method you can speak from the most easily.
- *Practice your talk,* when it is ready. This is paramount. Practice again and again (at least ten times, for good measure). Do this by yourself and in front of colleagues, until it becomes fluid. Practice breeds confidence and a sense of control over the material. There is no substitute for this. Think of what it takes to perform a piece of music in front of an audience.

When you practice, pace yourself. A good rate of speed is about 90–120 words per minute (average 100). This means for a 20-minute talk, about 2,000 words; for a 30-minute talk, about 2,800 words (providing for pauses, etc.); and for an hour's talk, about 5,500 words. These rates are significantly slower than conversational speech, and they need to be. Measured speech allows the audience to keep up with you, and it gives you the appearance of control and fluidity.

STAGE FRIGHT: A UNIVERSAL

Please don't doubt, even for a moment, that all of us are subject to stage fright. This is natural no matter how many times we stand before the lectern, settle our notes, and gaze out into a sea (or pond) of faces. Speaking is an intense experience; one feels exposed to the ferocious scrutiny of strangers.

Yet, as veteran presenters can tell you, your listeners are far more likely to sit before you in sympathy. Many or most have been there, and know what it feels like. If asked, they would readily admit appreciation of your efforts to share your work. It is thus with these perceptions—which, in effect, validate the very reasons for giving oral presentations—that you yourself should begin.

Experts on public speaking will tell you that stage fright has a positive side. It is not merely fear, but also excitement. We wouldn't feel nervous if we didn't care about our performance or the content of what we're about to offer. This suggests that you might try to aim your nervous energy in positive directions. Breathe slowly; think of your speech as an act of intelligence and bravery (which it is). Visualize yourself doing well, moving through the material easily. Hear your voice saying the opening sentences of your talk, then going on from there.

The spoken word is the music of knowledge: momentary, personal, expressive. No matter how scripted it might be, how calm and replete with specialized terms or tones, it emerges from you, the individual speaker.

STRUCTURE AND FLOW

When you design your talk, try to establish a fluid logic between your main points. It can help to write out the primary topics you think you might want to cover, or lay out the visuals you may want to include, then search for an order among them. Don't be afraid to do this at an early stage. Try to experiment: include more topics or visuals than you might need, move pieces around, take some out, see if a particular sequence appears.

In doing this, think of someone in the audience who will be taking notes on what you say. They should be able to put down your major points, in order, without difficulty. Your presentation, that is, should include discussion of its own structure. It should cover the following:

- Why you did this work (emphasis on "this")
- How you did it (what tools, techniques, approaches you used)
- What your findings were (your results)
- What you have concluded from it (what does it all mean?)

If you are able to cover these areas, in order, within the time allotted, and without causing your listeners to go unconscious, you have probably done a good job.

How to begin? In most effective presentations, a speaker will start with a word graphic showing his or her name, affiliation, and the title of the talk. This may seem unnecessary, given that the talk has been listed and an abstract probably made available. But human reality requests certain considerations. During the first few minutes, the audience will be adjusting from the previous talk, settling into their seats, still leaving or arriving, sipping coffee. A moment or two of introduction is therefore needed; it will also help get you, the speaker, acclimated and focused.

Next, it is a good idea to show the audience how your talk is organized. This can be done using a second word graphic or two covering an outline of the major topics. Introduce this material with a phrase such as, "today we will be talking about" or "in this presentation, I'd like to cover the following." Your graphic, meanwhile, can provide a list of statements devoted to each of the above questions (why you did this work, etc.), or a general overview (three or four areas to be discussed). In all cases, it works very well to present these points at the beginning and then return to each one, sequentially, later in the talk, when you treat it in detail.

Think of your talk, therefore, as divided into sections. You might begin each new section by repeating your outline and highlighting the portion that will be discussed next. Cue these transitions verbally, too, with such phrases as, "let's turn now to" or "having looked at . . . , we can now examine." Finish up with a review of your major points, followed by a set of conclusions. Keep these as simple and direct as you can, as they will be the last things your audience will see and hear. Between your opening and closing, you want to move your listeners through your story as smoothly as you can. Try and order your visuals so that the audience can easily follow your logic from one to the next. Think of including a few "relief" images, that is, those that are more purely visual (photographs, models, maps) and might give your listeners a break from a series of text- or number-heavy graphics.

Finally, don't think of the time limit (10 or 20 minutes, usually) as a burden, but instead as a kind of liberation. Why? Consider the difference between writing a full-scale research paper and an abstractlike summary—which is easier? Would it really be less work to produce an hour-long speech on your topic? Working within the given time limit gives you the freedom to hit the high points and do no more. It releases you from the bond of having to prove every generalization or important statement; you need support only those you feel are most central to your story. A talk, moreover, gives you the chance to tell your tale according to the simplest, most Aristotelian of formulas, with a beginning, middle, and end, and a flow that follows the basic scheme of "I did this because . . . using these tools . . . and having found these results . . . made these conclusions."

USING VISUALS: REALITIES AND TECHNIQUES

Visuals in an oral presentation are far more than mere props or aids. They are very often the main bearers of the message. Many scientists therefore plan their talks on the basis of the images they intend to show. Whether you adopt this approach or not, it is important, very important, to plot out visuals (or type of visuals) at the same time that you plot out the structure and organization of your talk.

A proven technique for integrating graphics and speaking is to storyboard your talk. This involves actually mapping out the presentation, image by image, whether on paper or by computer (using presentation software). Common methods use one sheet of paper per visual, with the specific graphic pasted (or described) in the upper half and notes for speaking underneath. This has the advantage of allowing you to flip through your talk easily, noting any changes in organization that might be needed.

These days, computer software packages give you the ability to do all of this on screen (including laying out your material all at once). Some scientists find this new capability extremely useful; others prefer to stick with the older methods of paper and pen. One benefit of the computer approach is that you can create the actual visuals online, while you storyboard. You can also generate smaller thumbnail versions for easy use. Presentation software can produce a rehearsal slide show for you and makes it very simple to reorder, delete, or add material at any stage. This can really be a help, and can speed up your preparation significantly.

Another benefit of going online is that some graphics already in digital form, that is, that you've generated during your research, can be imported into a presentation, and then adapted into a slide or multimedia image. Here, however, caution is advised: only the very simplest illustrations may be suitable for direct use, and even then rarely without *some* modification (see chapter 9, especially figure 9.10 and related discussion, for an example). More often, you'll need to simplify and streamline your figure a good deal—thicken lines, enlarge text and data points, delete and trim material—so that it can be grasped within a very short time (ten seconds is not unusual).

GENERATING VISUALS: OPTIONS TODAY

What options exist for presenting your images? There are basically three: overhead transparencies, slides, and computer-driven displays.

The first of these was once very widely used, but is now passing out of fashion, except in more informal circles. Slides are most popular, but are being gradually replaced by electronic presentations.

Transparencies have certain advantages. They are cheap to make, require simple equipment, can be shown in either light or dark settings, and can be written on. Their disadvantages are also well-known: they encourage copying of material from the published literature (i.e. material designed for print, not presentations); they can be sloppy; they are clumsy to handle; and they offer little by way of aesthetic attractiveness. In many areas of science, overheads remain in use for small gatherings, but have been largely replaced by slides at large formal gatherings, such as annual meetings, conventions, and symposia.

Slides, though static, allow you to show a much larger array of visual material than do overheads. Presentation software has helped this medium become more colorful, visually diverse, and effective. The aesthetic richness of the spectrum has brightened science in many quarters, but none more than presentations. Even charts, tables, and word graphics are now regularly presented in multiple colors.

Nowadays, moreover, it is becoming more frequent at scientific meetings to use two screens and two slide projectors simultaneously during each talk, allowing (forcing?) the speaker to show even more material, without having to constantly "flip" the view. This obviously presents the opportunity to also bewilder or overwhelm an audience, when *too* much is delivered. I would ask that if you're using two screens, consider that the audience will usually focus on the left—we write from left to right—so put your most crucial data, information, and images here.

Such decisions, as well as the general trend toward more color, has been greatly aided by the advent of computer software designed for generating presentations. Such software can import almost any type of digital file—charts, scanned photographs, simplified maps, models—to a mock-up slide, on which you can then add words or highlights. Color schemes come ready-made and, in most cases, seem intelligent enough. Most software packages of this type are quite easy to use; they require little learning time and offer huge benefits in terms of generating and revising (very important) your slides quickly and efficiently. There are even programs and equipment (not expensive) that permit you to create the physical slides yourself. Alternatively, you can save a presentation to disk and have slides made from the files. If you wish to save money, you can simply photograph a presentation directly from a computer screen.

More efficient than these methods, and increasingly more popular, is to store digital slides on a compact disc or portable computer and bring these to a meeting. Projectors and related equipment are now regularly provided to handle such material, which has the considerable advantage of doing away with the need for physical slides. Such presentations can also be easily updated or modified for later presentation. Moreover, digital slides can be enhanced by various "special effects," such as fade-in transitions (many other types are available) and limited animations.

Thus, while old-style (physical) slide presentations remain popular at scientific meetings, the future is already lapping at the shores of some disciplines, as electronic multimedia displays advance. This is not merely a matter of the eagerness to use new technology. Physical slides are static; electronic display makes them dynamic, able to be manipulated. Using a portable computer and projector, you can create a large number of effects that not only provide enhanced visual interest but also allow you to make excellent transitions and, more importantly, the adding of much new content. For example, you can add or remove highlights or inserts in real time; pull out (magnify) different portions of an image; construct graphs on the spot; present and rotate three dimensional views; show brief video clips or animations; and so on.

Electronic presentations are the tide of the future, yet they have certain limitations today. They are more complex and time-consuming to build. There is a steeper learning curve. They are also intricate to set up, physically speaking, often involving a large set of wires and connections that can be distracting (this will surely change, when wireless hookups become the norm). These factors combine to bring a degree of uncertainty to the actual performance itself. At a recent conference I attended on new and emerging technologies in the energy industry, for example, it was hard not to notice that over half of the talks had to be stopped and restarted several times because of problems with the computer-driven displays. The irony separating message (technology is the future . . .) from medium (but fails us in the present) was not lost on the audience. Watching the presenter and technicians fiddle with the equipment reminded me, time and again, of the days in science class when a film strip would start to jump or go black, and the AV (audiovisual) student would have to work feverishly to get it going again. Impatience and a lapse in interest were the price paid (as well as spitballs and note passing). Those too eager to use technology seem fated to repeat its past, whether as farce or tragedy.

Consider, too, that such technology may be more appropriate to certain types of presentations than to others. I've seen many examples where little more was done with them than to add highlights to word graphics or animate one or two images that would have worked just fine with a static display and laser pointer. While there is advantage to trying out new technology, visiting such attempts upon the heads of live audiences may prove to be a poor choice. Special effects can be distracting, flashy, and suspect. Keep a critical eye open: look for both poor and excellent examples and note them down for future use.

There can be little doubt that electronic displays will one day come to dominate professional talks. The technology may not be wholly there at present, but it soon will be. Yet with this new tool come new responsibilities for making aesthetic judgments about how to treat different types of content.

DESIGNING VISUALS: SOME PRACTICAL POINTS

Keep track of the things you like and dislike about other scientists' visual materials in talks, what you find effective and what strikes you as confusing or ill-chosen. Do this in a general enough fashion so that you can learn from it and use it to strengthen your own presentations. For example, make note of what it is that renders a particular chart or table easy to read (Well-sized type? Good contrast? Clear design?), or, on the other hand, what renders it incomprehensible (Too much information? Type too small? Poor use of color?). You might see a very nice display of routine information or an innovative arrangement of data or graphic material—write this down, if you can, for reference. If possible, ask the presenter for copies—don't be shy; this is a great opportunity to help improve your own work, possibly to make a new contact, and to offer a compliment that may be returned to you in the future.

In some cases, the proceedings from a symposium, conference, or seminar will include printouts of slides shown during the talks. Such proceedings, therefore, can be a gold mine of visual material for you to evaluate and learn from. If at all possible, collect the best images in a file, along with the textual models you may have chosen. Review these before designing any future talks, to keep their lessons fresh.

Aside from what you will learn in this manner (a great deal!), here are some pointers to keep in mind.

Keep your graphics simple, as simple as you possibly can. I can't emphasize this enough—visuals with too many things to look at, too much text, too many patterns, too many lines, are one of the most frequent problems in scientific talks. Remember that your audience will be looking at each visual for *a minute or less* (usually less), a very short amount of time. As much as a third of this time will involve tuning in to the new information shown. Therefore be considerate: give them something they can easily make sense of.

Design any text so that it can be deciphered clearly, from a good distance away. This includes titles, labels, scales on graphs, tables, and so forth. Bad text is also extremely common and very aggravating, showing a lack of forethought. Good text expresses sensitivity to the situation: viewers will see that you are thinking of them.

Similarly, all content must be distinctly visible. Use brighter colors for more important information, duller colors for background. Think about using the brightest colors for anything you want to highlight or make stand out. If the information is mainly visual and not textual (e.g. patterns), reduce all text to an absolute minimum. Know that people will try to read any text you put up, so if it is not entirely necessary, it will end up being distracting.

When you put up a word graphic, try not to duplicate it exactly in your own speech. Choose the most important phrase in each listed point to recite (this can provide good emphasis), but enclose it in remarks that differ from those on the screen. Add a bit of commentary, if needed, or explanation.

Good slides can be used again and again, like a resource. They can be reused either in their original form or in modified fashion for successive talks about ongoing work. Thus, time spent on designing them well is often time saved in the long run.

SHOULD YOU READ A WRITTEN LECTURE?

Should you read a written lecture? There is no single answer to this question. While many scientists would likely sacrifice their first-born child rather than see this become accepted practice, many others do not feel so strongly. In some fields, true enough, it is strongly discouraged, even punished. In others, it is grudgingly allowed; in still others, however, it is fairly common and viewed with a mixture of forbearance and unconcern. In the geosciences, for example, all of these attitudes exist and are divided among various subdisciplines. Academic organizations tend to prefer nonwritten talks; those with some connection to industry are more open to presenters' reading their paper.

Objectively, there is no reason why reading your paper should be a forbidden means of communication. But—and this is a large qualifier—there are a number of trade-offs and risks involved in proceeding this way. First, reading a paper out loud can be a deadly affair for your audience, *unless* you have taken the time and intelligence to write it with many colloquial, conversational aspects. The normal scientific paper is about as far from ready material for a speech as a mathematical proof. Reading such a paper is guaranteed to bore and stun and thus offend. Second, when you read a paper, your presence for the audience is significantly reduced: your eyes and voice are directed downward. Unless you are able to lift yourself by looking up frequently and projecting well, you will tend to subside into the page (possibly for emotional reasons) and may even disappear altogether. Thus, to project an equal degree of personality, a reader must be even *more* skilled than the speaker who performs extemporaneously. Third, reading a presentation while trying to discuss visuals can be awkward, if the verbal choreography isn't well-rehearsed. All of these realities mean that to deliver an effective talk in this manner requires more work than an extemporaneous speech.

Again, there are no ultimate rules, no moral absolutes, in this area. If your field allows for such talks (as mine does), then they are a practical option, as long as you understand and adjust to the trade-offs involved. Perhaps the most fundamental argument against reading a talk is the truth that the audience has come from near and far to listen to, and interact with, the human side of science—that is, individual people, not narrating machines. A fallible, slightly anxious speaker, who may stumble now and then, but who speaks to the audience, is much more interesting than a smooth but monotonic reciting device. Thus, if you do decide to read your paper, begin and end by not doing so. Introduce your talk, and close it, by speaking directly to your listeners. Let them know that you are there, not as a hologram, but as flesh and living voice.

LECTURING TO THE PUBLIC: A FEW POINTS

Here let me offer just a few guiding suggestions. If you are fortunate enough to be asked to make public lectures, whether to local societies, university gatherings, or teeming halls of interested laity, you will face a somewhat different series of challenges, and rewards. Public speaking for scientists is once again, as it was in the 19th century, a booming industry. There is considerable interest in science among general and educated audiences today, and the result is an opportunity for researchers to become eloquent spokespeople for their (inevitably glorious) discipline.

Consider, therefore, these suggestions to make your lecture effective.

Prepare

Know who your audience is, how large it will be. Keep these things in mind as you design your lecture. For larger groups, you'll need to be formal but entertaining. For smaller audiences, a more relaxed, conversational bearing will be in order.

Know where you will be speaking, what technology will be available. Make sure well ahead of time that the facility will have what you need, or that you design your talk with the facility in mind. Don't assume there will be projectors, blackboards, carousels, pointers, water, whatever. Find out or request. Moreover, if you will be showing slides, always bring your own carousel and your own laser pointer—it doesn't hurt to be prepared for any emergency.

Consider: the actual goal of public speaking is to *connect* with your audience, not to overwhelm or impress them with your depth of knowledge and imperial bearing. If you truly wish to represent your work and field gallantly, in memorable fashion, you must inform your listeners and give them pleasure as well.

Design

Begin thinking about your talk at an early stage. Prepare a mental "net" for catching interesting material. Outline your topic in general terms first, then keep your eyes and ears open for any facts, statistics, visuals, anecdotes, historical information, and so forth that might be good to include. If appropriate, cast your net widely: see how your topic might have been covered in newspapers, on television, in the movies, in other more popular or public venues. Such coverage can provide valuable material for an opening, closing, a point of discussion, even a source of amusement.

If possible, come up with a central theme for your lecture. This can be a simple as, "Where we are today in astrophysics," or as complex as, "What is a genome?" Try not to be too cute: puns based on movie or television program titles can trivialize a topic and make you sound condescending.

If you are seeking to bring an audience up-to-date on discoveries in your field, consider posing your lecture in the form of an actual story, a narration on how certain advances came about. Start by discussing the crucial scientific questions that remained unanswered (or unanswerable, at the time), and go on to show how new approaches, methods, and ways of thinking were conceived and carried out. You might end such a presentation by talking about the new questions this research has produced, thus forming a neat cycle (and an image of how science sometimes works).

Delivery

Begin with a strong opening—something that grabs, that can spark or deepen curiosity. You might choose for this purpose a remarkable fact, a striking anecdote, a quote from one or another famous source, a beautiful visual. Pick up on this opening later in the talk and, if possible, return to it at the end.

Smile occasionally as you talk. Make eye contact with the audience: focus now and then on a few listeners who nod back at you or otherwise indicate connection. Move around a bit; use your hands now and then. Try to stand in a relaxed posture. All of these aspects of your delivery will keep the audience attached to you and what you're saying. They like to see that you are human and alive.

Humor lubricates a message, relaxes an audience, lightens the atmosphere. One bad joke, however, will stiffen everything, grind the machinery to a halt. Think carefully about what types of humor you might employ: try them out on others before you dare do so on a room full of strangers.

Think about using questions as a way to make transitions. These can be rhetorical: "Now, where did we go from here?" More suggestive: "How were we to interpret *this* result?" A directive: "The next problem we encountered was, What to do with all this magnetic data?" Teasing: "But if that were the case, why study this species at all?" And so forth.

Possibilities for adding "color" to any talk: anecdotes or stories (make these relevant to your topic!); historical information, including images; autobiographical details (again, make them relevant); quotations from literary and/or scientific sources (use sparingly); cartoons (always check to see whether permission is needed); demonstration (use of props, drawing on blackboard, or simply using hands); multimedia (images that move, use of sound, video clips, etc.).

Try to think of questions that might come up for the audience during your talk and answer them as you go along. You might even pose some of these yourself, as part of your delivery—a technique almost sure to endear you to your listeners.

Close with a strong, well-phrased ending. If possible, return to your beginning and use it to create some sort of closure. Speak of what directions future research might take, what questions remain unanswered. Leave the audience with something solid.

If there is a question-and-answer session at the end, make sure you repeat each question out loud, so that everyone hears. Try to keep your answers fairly short. Don't let any one question monopolize your time; if you have an insistent questioner, ask them to come up and see you afterward.

Attitude

Consider that, in some respects, you may be donning the mantle of "the Scientist." This means that you have instant credibility, but also that you may need to convince your audience you are fully human and capable of saying things in interesting ways.

Beware the temptation to condescend. This can come in subtle forms, or not so subtle ones. Avoid too many "Hollywood touches," such as spectacular visuals (one or two is OK; a dozen is not), amazing facts, "mysteries of nature," and the like. Also try not to pander to the lowest common denominator by falling back on clichés ("a hundred million of these could fit on the head of a pin") or simple-minded analogies—especially those of the kitchen ("in their early history, planets were like hot pancake batter") and, above all, the sports scene ("this was a slam-dunk result"). A few such touches are fine, even expected. But a consistent reliance on them will make it look like you are "dumbing down" your material. If at all possible, be more thoughtful about the imagery you use—it will make you a much better (and more quotable) spokesperson for your subject and field.

CHAPTER FOURTEEN

The Online World

Using the Internet

A NEW MEDIUM WITH NEW MESSAGES

Let us begin again with history. It shows us two things: first, that the forms for communicating knowledge have evolved continually from the birth of writing, about 3200 B.C., to the present; second, that there have been times when specific new media have appeared—the scroll, the codex, the printed page, and now electronic display—and changed profoundly how people record and exchange learning. These realities suggest that the Internet will continue to develop for some time, and it will do this alongside existing forms of exchange, not as a full-fledged replacement of them.

Many people, scientists included, have claimed the Internet as the next great revolution in human communication. Whether this is true cannot yet be said with certainty; we are too much in the midst of it all. It may turn out that the online world constitutes a new direction in a more long-term move from printed to electronic forms (telegraph, telephone, sound recording, video, computer). But there can be no question that the Internet is an innovation of enormous impact and thus import—to science, perhaps most of all.

Why science, above all? There are several reasons. One is the enhanced contact among researchers throughout the world, enabling new collaborations and the transfer of information to a much wider audience. What has been termed "the invisible college" of science has therefore expanded tremendously, with many productive results for research. Another reason is the Internet's ability to *distribute* research in almost any form, including text, video, audio, and any type of image, fixed or animated. This is a crucial factor, not to be underestimated: the role of various media in science is more important than ever, and has been greatly expanded by the advent of digital ways of embodying knowledge. Various types of complex visualization, for example, are now at the heart of many fields.

Online publication, meanwhile, can handle enormous amounts of data—raw and analytical—without increasing physical storage space, but with greatly enhanced searching and indexing. It can shorten or even eliminate the lag time between submission and release, typical of print journals. It may also prove helpful in dealing with the "serials crisis" now besetting research institutions throughout the globe, a crisis of costs (escalating subscription rates, growth in the number of journals, reduced budgets) that has forced libraries to be ever more selective and thus incomplete in their collections. Finally, the Internet may well prove able to give rise to *new and unpredictable* forms of science altogether. Versatile, fast, expansive, and cutting edge: the Internet is an exciting set of opportunities for researchers to explore.

ISSUES FOR ONLINE SCIENCE

These same powers have resulted in new problems and issues, too. Most basic of all, perhaps, are concerns over copyright and proprietary information. The Internet allows for instant duplication and nearly infinite distribution of any data or written material in digital form. Moreover, the fluidity of Internet communication renders more tenuous the very definition of a legitimate scientific publication. Legitimacy, after all, has traditionally come from certain institutional factors: peer review, editorial control, document stability, guarantee of persistence, and copyright. All of these emerged under the rule of print; most have been in the immediate control of editors and publishers (for better or for worse). Some (peer review and editorial control) have translated to the Internet without too much problem where online versions of print journals are concerned. But these represent only a small percentage of the total new range in scientific publication. There are also new electronic-only journals, preprint archives, personal Web pages, research newsletters, e-mail documents, and other opportunities for publishing. Such material does not yet have an agreed-upon status. In some cases, it counts as sanctioned research; in others, no consensus yet exists.

Documents on the Internet are not necessarily stable like those in print. This is because the Internet itself is not stable. Any text can be revised over time, updated with new data, expanded through the inclusion of reader's comments, taken apart and reassembled. How, in such cases, is one to determine an official version, for example, for citation purposes? Then there is the very important problem of archiving—keeping scientific e-documents available in perpetuity, at a specific location (e.g. Web address). Should we, as scientist-authors, be able to retain sufficient rights so that we can include our own articles on our own Web pages, or others of our choice? Such are among the many issues that need resolution.[11]

But if the fluidity of the online world has given new uncertainties to scientific publishing, it has also added strength to a traditional reality. The Internet has deepened, not weakened, the centrality of the written word in modern intellectual society. There is now a vast and growing array of new outlets and forms of knowledge, true enough. But this knowledge remains utterly dependent on written language, for this is primarily how society continues to embody intellectual work. The Internet may be digital, electronic, and "nonlinear"—but it depends utterly on the inscribed message. The conclusion to be reached from this is clear: if the pace of science is increasing, so too are the demands on scientists to possess adequate writing skills.

STATUS OF THE ART: SOME MAJOR TRENDS

To try to describe the status of science on the Internet, from the writer's point of view, is to take aim at a moving target. Moreover, it is a target that changes both shape and speed with each passing year. Safe to say, however, at the beginning of the new millennium, that online science is moving rapidly toward the center of communication in every major field: physics, chemistry, biology, geology, astronomy—you name it. Any scientist who is not ready to utilize this medium will therefore soon be left behind in significant ways.

What are some of the major trends in Internet science? E-mail is now a primary mode of both informal and formal contact between researchers and others involved in scientific enterprises. Immense amounts of raw and interpreted data are now available in online archives from academic, governmental, and industry sources. Unprecedented bibliographic resources—some with the ability to locate, retrieve, and deliver

11. All of these issues have been extensively discussed in both online and print literature. One of the most thorough overviews can be found at the Web site of the International Council for Science (http://associnst.ox.ac.uk/icsuino/). See, especially, the proceedings of the various listed meetings and the online booklet *Guidelines for Scientific Publishing*, available in both HTML and PDF formats.

not just citations but abstracts and full-text articles from a vast array of periodicals—have been created and made available. Nearly every major journal now has its online version. Many of these, in fact, are more extensive, interesting, and informative than their print cousins. Large portions of the scientific literature have been transferred to the Internet, often with varying degrees of success. Once-firm boundaries between primary and secondary have dissolved, as journals have taken up aspects of newsletters and vice versa; as data archives offer reviews of the current literature; and as the Web sites of individual researchers include everything from published and preprint articles to course outlines and opinion pieces.

Publishing material on the Internet, especially the Web, is not yet cheap. One of the largest expenses in print publication is labor—people doing the work of copyediting, proofreading, formatting, and other tasks associated with quality control. This work is not eliminated by the online universe. Moreover, there are new costs in preparing e-documents, whether this involves designing Web pages, coding text, or scanning images. Then there is the expense of putting material on the Internet through a server, and keeping it there indefinitely. On the other side, potential readers must have access to a recently built computer with an Internet connection and the proper software—something that is very far from universal in the scientific world beyond the edges of the major industrialized nations.

There are, moreover, certain signs that Internet publishing is still at an early stage. Much of this is related to technology, but not all. Downloading data and articles can be a slow, laborious process, due to large file sizes, restricted modem speeds, and heavy site use. There are still a number of different formats in which text, numerical information, and images may occur. The lack of any standard in this area means that information does not always translate well, or at all, between computer systems. This can require scientists and universities to purchase redundant software in order to ensure the widest access to needed data.

Things also tend to mutate on the Internet very quickly. Web addresses, new journals, online articles, and much else can shift or vanish without warning, due to ordinary institutional circumstances, such as loss of funding or personnel changes. This leads to the phenomenon of the dead link—the hyperlink, either within a document or in a listing of sites, that goes nowhere, for example, to a "Web page cannot be found" type of message. This is surely one of the most frustrating parts of the online experience, and one of the most common. Finally (but this list is not complete), there is the problem of searching—Internet searches still comprise a very rough art at best and often require trial-and-error iterations.

As a new mode of communication, the Internet may well be "revolutionary," but like all revolutions, it is messy, changeable, and beyond the control of any single entity. Scientists (like everyone else) should be aware that however smooth and magical it may seem, the Internet is not really virtual—it is the expression of people performing certain tasks, with certain equipment, under certain conditions. It is no more "without walls" than a laboratory.

EXISTING RESOURCES: WHAT THERE IS AND HOW TO SEARCH FOR IT

I have mentioned several sources where scientific information is published on the Internet. There are others, too. A basic list, roughly in order of importance to writers, includes the following:

- Online journals, newsletters, magazines, publishers
- Preprint archives (papers written but not yet formally published)
- Major bibliographic resources
- Professional society and association sites
- University department sites
- Library sites (academic, governmental, independent)
- Data archives (domestic and international)
- Personal Web pages of scientists
- Government agency and program sites (NASA, NOAA, DOE)
- Research program sites
- Industry sites (individual companies, consortia, etc.)
- Research institutes
- Local scientific society sites
- Observatories
- Image catalogues and archives

This list offers some idea of the scope that now exists. It is no longer possible, in most fields, to comprehensively survey relevant sites, as there are simply too many. Moreover, these resources are extremely variable, both in content and in quality: some provide raw data by the hectare, others little more than a visual brochure for a particular program, publisher, or department. Unfortunately, there is no simple way to separate shells from seeds in this endless garden of seeming delights.

Scientists need to be aware of several basic ways to look for specific material. For literature searches, and for tracking down abstracts and even full-text articles, you're best bet by far is to use one of the major online bibliographic resources—nearly every field now has one. These have quickly become *invaluable* reference and research tools, offering unprecedented access to an enormous spectrum of information. Some have been around for years—Agricola (journal articles and book chapters acquired by the U.S. National Agricultural Library), BIOSIS (comprehensive coverage of international life sciences literature), MEDLINE (coverage of more than 3,900 journals in biomedicine), Geobase (citations and abstracts in geographic and geologic sciences), and Georef (international geologic sciences literature). But many others, specific to particular fields, research areas, and specialty topics have sprung up. Here are just a few examples: AIDSLINE (journal articles, theses, technical reports, meeting abstracts, books, and audiovisuals on AIDS and related topics); Physical Review Online Archive (American Physical Society's effort to put all APS journal material online, covering the years 1893 to the present); Aquatic Sciences and Fisheries Abstracts, ASFA (covering every aspect of aquatic science), PsycINFO (citations, summaries of journal articles, book chapters, books, reports, etc. in psychology-related areas); TOXLINE (toxicological, pharmacological, biochemical, and physiological effects of drugs and chemicals). Readers should note that the specific contents and Web addresses for these resources may change over time. As pointed out, the Internet remains a fast-changing medium, subject to many influences.

Most universities, research institutes, and companies performing frontline investigation now subscribe to several of these resources. In the case of major universities, the list is likely to be large indeed, and growing. Every scientist should be aware of what is available to her or him, both in her or his chosen field and related fields as well. As always, explore, explore. Learn how to use these resources; it will grant you new powers and efficiency.

For other material on the Internet, two basic options are available. You can first try one of the mass-market search engines (e.g. Infoseek, Google, Yahoo, Lycos, Searchalot, Alta Vista, HotBot, etc.). Chances are, however, that because of how these work—most look for your entered terms in any text on any Web page—you are likely to wind up with a motley selection. Putting a phrase in quotations ("molecular modeling software") will tell these engines to look for your topic exactly as written; but here, too, you may end up with pages that only mention these words, rather than actually provide the needed material.

A second, and generally far more effective, approach is to find a type of "gateway" or "portal" site in your particular field—a site, that is, containing a large list of links to more specific sites—and to begin your real search here, bypassing the majority of irrelevant material.

Such gateways take a variety of forms. The simplest can be found under headings like "Science" that appear on the Web pages of the major search engines: clicking here will bring you to a roll call of separate fields, then subfields, then individual resources. Other, more "professional" portals are likely to exist on the home pages of university departments in your field, professional associations (e.g. American Chemical Society), and major research libraries, such as the National Library of Medicine in the United States—any of these will be a good place to start. Similarly, the Web pages of relevant government agencies and research programs, and (do not overlook this) the personal Web pages of influential researchers, will probably include an inventory of selected links to useful material. Finally, indexes—stand-alone sites that offer sometimes huge lists of sites, hopefully annotated and grouped into categories—can be very helpful, once you've located them, explored what they have to offer, and know their strengths and weaknesses.

Possibly most helpful of all are what might be called general resource collections, a breed of Web site that, in time, may supersede all others in utility. This kind of site gathers together a wide variety of links related to exactly the type of information scientists want on a regular basis: research tools (databases, downloadable software, etc.); online journals; summaries of the current literature; reviews of new books, software, and labware; news and commentary related to the profession; conference reports (including audio recordings); job postings; classifieds; and more. This type of gateway is obviously an excellent idea, and it is something that only the Internet makes possible. Depending on who maintains the site, access can be free or available through subscription. Those put up by researchers or university departments tend to be free. Sites maintained by major publishing houses, on the other hand, are not cheap but do provide a lot of content (a good example, for the biomedical field, is BioMedNet, at www.bmn.com, put up by Elsevier Science Ltd.). As always, it pays to investigate what is out there and to be selective.

In every case, think of your search as a type of bibliographic exercise, requiring both cunning and endurance. Looking for data can be akin to finding a book whose title you do not yet know. As with library research, it usually helps to locate the series of "shelves" (resource sites) that hold books in your area of interest—and then to go through them, possibly one at a time, until you find what you need. Be prepared: certain books may be "out" or "missing"—the phenomenon of the dead link continues to abound. If an index or gateway leads you here fairly often (say 20% of the time), abandon it and go elsewhere, for it probably isn't updated regularly. Be aware, too, that many resource lists are now *huge*. Thus, your quest may end up a small project of its own—but with very worthwhile results.

Online searching therefore involves exploration, with all its attendant difficulties and discoveries. Be prepared for frustration—but also for unexpected riches. Above all, do not anticipate instant results. When it comes to Internet science, patience is not merely a virtue but a required tool. There is no substitute for personally learning the lay of this ever changing and evolving landscape.

E-MAIL: BENEFITS AND CAUTIONS

E-mail is truly a new and very flexible mode of communication, and an enormously powerful one for researchers. It has probably had more immediate effects upon contemporary science than any other aspect of the Internet.

E-mail is at once extremely easy to use, instantaneous, and potentially global. Its ability to establish new contacts, and to maintain existing ones, is unparalleled. Its power to exchange information among individuals, groups, and institutions is bounded by neither time nor place. E-mail encompasses both informal and formal types of expression—all levels of written communication in fact, as well as nonverbal forms. If it has partly replaced the telephone and fax machine for quick contact, its ability to send attached files has made it a preferred method for submitting abstracts and articles to conferences, journals, and publishers, as well as a regular method for sending all manner of data files. For these main reasons—ease of use, global exchange, rapidity of transfer, and flexibility—e-mail is now the most widely employed form of communication among scientists generally.

There are several limitations to consider, however, and these are important, especially for writers. First, e-mail is not (yet) universal; one must obviously have access to the proper equipment and connections, and large portions of the scientific world still lack these. Second, the technology is not perfect, especially when it comes to transferring files. Depending on your Internet connection, even moderate-sized files (e.g. three to five megabytes) can take a long time (up to 20–30 minutes or more) to download, particularly during periods of high traffic. This will undoubtedly improve over time, with faster connections and broader bandwidth; but don't look for improvements to happen overnight.

Then there is the question of "personal style." Because it is so immediate, e-mail writing lends itself to great variations in personal tone. Some researchers choose to write a blunted, telegraphic prose; others use a friendly, conversational manner, as if speaking over a cocktail. Still others tend toward a tone similar to a formal letter. Most scientists probably alternate between these and other styles, depending on the occasion. As a result, certain mismatches can occur, for example when an affable message of several paragraphs is given a curt, one-line reply, unsigned. This sort of thing can cause bruised feelings, even problems of respect.

Therefore, two points need to be made. First, don't take offense automatically. An e-mail persona may reflect how its owner views the medium itself (e.g. as a simple messaging service rather than a chat opportunity), not you or your message. Second, and conversely, think before you write, and reread before you send: consider whether or not it's a good idea to answer that brief request for an abstract with an extended discussion of your recent fishing trip to Montana, complete with attached photographs (in color, of course).

There are, too, certain codes of e-mail behavior that you need to know. Various print and online guides exist in this area (e.g. www.emailhelp.com/etiquette.html). The most frequent problems involve use of capital letters. Putting a sentence or word in all caps is *not* the equivalent of adding an exclamation point: it is much closer to raising your voice, even yelling. Therefore, employ this with caution and emotional intelligence.

Above all, remember that e-mail is *not* conversation. Anything you put down is recorded, and as such, can be sent elsewhere, printed out, and extracted from the hard drive of your computer or that of your recipients, at a later date (erasing a message doesn't make it disappear). In short, e-mail is potentially public. Again, think before you write and send.

Partly because of these realities, e-mail is now covered under U.S. copyright law. All messages that originate within the United States are protected under this law, and anything you write can be kept confidential—*if* this is agreed upon beforehand by you and the recipient. Copying (quoting) small portions of any message is allowed only to the extent permitted by the fair-use provision of the copyright law (e.g. quotes from books in reviews). This means that posting an entire e-mail of another author without his or her consent constitutes a legal violation. Although, in my experience, this is done quite often without much thought, it can be subject to prosecution. Therefore, it's a good idea to ask before you share someone's comments or information with others.

In the end, the e-mail universe reflects the complexities of the social dimensions to scientific work. While many scientists (and scholars generally) treat it very loosely, this universe does depend on certain rules and conventions. It is *always* a good idea, therefore, to be aware of what you are saying, how you are saying it, and who you are saying it to. A number of researchers I know force this upon themselves by beginning a message with "Dear . . . ," thereby making it plain that they are not just chatting but producing actual hard copy.

E-JOURNALS IN SCIENCE

By far the most significant challenge to the older ivory-tower journal system, with its cumbersome lag time, is the peer-reviewed e-journal. This has rapidly and unevenly become a major medium for disseminating research throughout the scientific world. Only a few years before this book was written, there were probably less than 100 e-journals in total. Now there are thousands, with a growing number of e-only versions. Indeed, whatever new opportunities the Internet may provide, it is clear that the journal article, in something not too distant from its present form, will remain the nucleus of the *corpus scientia* for some time to come.

E-articles can be, and sometimes are, much more than their static print equivalents. They can contain hyperlinks to any cited literature, to authors' e-mail addresses, or to relevant online data sites. They can therefore render the reference list of old into a completely new and extremely useful portal. The e-paper can also include animated graphics, movies, and sound files, which is a great advantage to many fields where studies of natural motion, evolving states, or sophisticated modeling are important. E-articles can also be expanded to include reader commentary and debate (a number of e-journals openly encourage this). All these possibilities are currently being used and explored.

Any e-journal, then, can encompass such possibilities in multiple form. This type of publication is thus far more than a journal per se; it is something closer to a type of virtual research forum. The better funded and more stable of these now commonly have most or all of their back issues online and offer search functions for the entire archive. They provide links to many related e-journals and other resources.

Many allow users the option of subscribing to the journal itself (most often at a cost), to a weekly or monthly e-mail message containing tables of contents (usually free), or to a periodic listing of abstracts (free or at reduced cost). All this can be very helpful, though it doesn't quite replace in ambience that trip to the library reading room.

In specific format and style, e-journals vary more than their printed relatives. Many now offer articles in one or both of two text formats, which have become the most popular in use. These are HTML (HyperText Markup Language), the coded language of the Web, allowing for the inclusion of hyperlinks and offering maximum flexibility in terms of text, but limited ability to handle mathematical symbols and formulas; and PDF (Portable Document Format), which resembles pages of a printed document and is largely "dead text," easy and familiar to read but clumsy to handle and unable to take advantage of the Web's most interesting and advanced features. Each format has its pluses and minuses. HTML can be decoded by any Web browser, but converting a digital file to this format can be a labor-intensive, time-consuming process. PDF documents are often large, require additional software to read (freely available, however, from www.adobe.com), and cannot (yet) include hyperlinks, animations, or other sophisticated aspects.

The great majority of e-journals will tell you up front what they make available for free and what they don't. Most offer open access to their tables of contents and abstracts. Full-text documents, however, are usually restricted to subscribers—who have paid for print versions or print and online versions together—though a sample issue or two may be available for viewing. Online journals with free access to all content do exist and have become more numerous. Don't assume, out of hand, that any of these are less worthy than electronic versions of print journals. Many are now subjected to the same rigors of peer review and quality control, so that the level of scientific data, analysis, and text are all high. The principal obstacles to the success of any such open-access e-journal have to do with *persistence*—who will continue to pay for it?—and *acceptance*—who (how many) will read it, submit to it, cite articles from it.

One of the great hopes frequently expressed on behalf of the e-journal is that it will eventually solve the "serials crisis," in part by transferring scientific publication back into the hands of learned societies and individual scientists. As it stands, printed scientific periodicals are overwhelmingly in the control of a few giant publishing houses, who frequently charge very high prices ($500–$2,500/year) for material that is essentially submitted and reviewed for free. Many scientists have become disturbed at the thought of having to buy back their own work at such rates. When joined with the sentiments of librarians, such perceptions have tended to seek alleviation in the many prophecies attached to Web publishing—those, for example, that promise scientists can take back control and offer journals, archives, and other data at very low cost. Experiments along these lines are ongoing; some have been successful, some not.

The bottom line is that nothing on the Internet is ever free or wholly automatic. Most of the labor needed to publish any article in print still applies to Web journals—especially those that are peer-reviewed and editorially quality-controlled—plus there are the new tasks associated with Web page design, archiving, updating, and more. The visual immediacy of the e-journal erases most traces of the work and expenditure required to make it a reality. But scientists shouldn't be fooled: publishing on the Internet requires support from somewhere. This does *not* mean that online publishing can't solve the serials crisis—on the contrary, it is clearly the greatest hope in this area. It *does* mean, however, that solutions will be neither immediate nor without cost, but will require cooperation, inventiveness, and commitment.

As writers and potential writers, scientists should take whatever time is needed to investigate e-journals in their field and to get an idea of their current status. These journals clearly form a crucial publishing element in the future of professional science. As more and more of them become fully accepted in the existing system of status and reward, and as they succeed in modifying this system in unforeseen ways, science will grow and progress, just as it did with the advent of print.

THE PREPRINT ARCHIVE

Beginning in the early 1990s, a new type of archive was pioneered at the Los Alamos National Laboratory in New Mexico (it has since moved to Cornell University, at www.arxiv.org, and is supported by a cooperative agreement between the National Science Foundation and Cornell). Led by physicist Paul Ginsparg, a team of researchers built an online repository of preprint research articles, run by automated software that handles acceptance, searching, and retrieval. Articles represent the very latest work; they are complete but have not yet been accepted for publication in a standard journal. They can be submitted in several ways, including e-mail, and retrieved using different aspects of the Internet (Web, FTP, e-mail). The basic intent is to provide a forum for new work in as timely a manner as possible. A particular preprint article remains in the archive until published, at which point it can either be removed or retained, at the author's discretion.

Though some concern was expressed early on about quality control, copyright, and other issues, within the physics community the idea caught fire. By 1996, the Los Alamos archive served over 35,000 users from more than 70 nations, with as many as 70,000 electronic transactions each day.[12] Even at that time, in other words, it was the most widely used forum on the globe for accessing research results. Today, the archive contains hundreds of thousands of articles, spanning a host of fields in physics and physics-related disciplines (biophysics, atmospheric and oceanic science, geophysics, chemical physics, medical physics, etc.), as well as mathematics and computer science. The concept of a preprint repository, in other words, has proven enormously attractive to scientists in many disciplines, and continues to expand its overall reach. Hundreds of universities worldwide now link to the Los Alamos site. There are also dozens of preprint bulletin boards, running the same software. Smaller archives, specific to individual fields, have been set up.

Yet the future of this type of archive, as a new standard, is far from certain. In some fields (notably chemistry), journal publishers have moved to restrain its use, mainly through the claim that appearance in this forum constitutes a type of publication, thereby rendering the material off limits for consideration in other outlets. To demand that submitted material be unpublished is standard among primary technical journals, of course. The question is what status should be accorded the electronic preprint: is it a true "publication" or not? This problem did not exist for physics, because of timing: the Los Alamos archive was set up in 1993, when very few physics journals existed online, and, in any case, is a government facility, supported by government funds.

If they become more widespread, preprint repositories could be an enormously powerful tool for science. For readers, the advantages are obvious: immediate, 24-hour access to the latest research, in familiar and usable formats, a year or more before such work can appear in print. For writers, meanwhile, the benefits are possibly even greater, including as they do all of these benefits, plus the ability to reach a much larger audience, on a timetable that the researcher can dictate. Scientists in general benefit greatly from having the very latest research at their

12. These figures are taken from the article "Winners and Losers in the Global Research Village" by Paul Ginsparg, available online at http://associnst.ox.ac.uk/icsuinfo/ginsparg.htm. The article was originally given as an address at a conference on Internet publishing held at UNESCO headquarters in Paris, 19–23 February 1996.

fingertips, as this keeps them at the forefront and prevents them from duplicating the work of others. In several ways, then, the preprint archive could provide an excellent way for scientists to take control over the availability of their work.

As always, however, there are limitations to consider. The most important is that, in the current reward system, preprint articles are unlikely to count toward tenure or other career advances, at least in any immediate way. To a large degree, a preprint archive relies upon the greater scientific community to make judgments about validity and value. But how are such articles used, exactly? How should they be cited, or should they be? Uncertainty in status makes such questions inevitable and difficult to answer, at least for the time being. Ironically, it may be the opposition from print journals that finally determines them to be true publications. In the end, researchers should be aware of this new forum and perhaps investigate its availability and status in their own field. As with most things related to the Internet, the landscape of the preprint archive will continue to evolve in coming years.

NEWSGROUPS: A MEANS OF INFORMAL PUBLICATION

Newsgroups form a separate portion of the Internet from e-mail and the Web. They employ a system known as UseNet (short for Users Network), which links subscribers into a single, defined community of active communicators.

A newsgroup is an online peer group. Members post queries for discussion, respond with answers or further questions, provide relevant articles, give Web addresses to appropriate sites, and do more—all within a specified subject area, whether this be a technical field, an interdisciplinary topic, or a particular phenomenon. Nearly every field and subfield in science now have one or more newsgroups. Some include thousands of subscribers; others, less than fifty.

When and if you join such a group and begin to participate in its discussions, it is important to keep in mind that anything you post goes to *all* subscribing members. It can therefore be printed out, copied, and redistributed, unless there are specific rules against such actions. To help ensure quality control, many newsgroups are moderated; some even have a formal review process for individual queries and responses. A fair number are unmoderated, however, so that there is no guarantee that postings are accurate or even relevant. In most cases, however, subscribers to scientific newsgroups are serious professionals. Nonetheless, as with e-mail, it is always a good idea to think before you write, and to read over what you've written before you send it out into the world.

Newsgroups can be an excellent forum for keeping in touch with latest developments, for trying out and exchanging new ideas, asking for feedback, searching for hard-to-find data, or being directed to useful Web sites. The major scientific disciplines (physics, biology, chemistry, geology, astronomy) all have dozens, even hundreds, of newsgroups dedicated to everything from the latest research findings to media coverage, new technology, education, and job openings. Anyone is free to inaugurate a newsgroup on a topic he or she thinks will be of interest to others. Lists of newsgroups in your area can be found either through a gateway site (e.g. they are often included on university-department or professional-association Web sites), through librarians at your institution, or in printed Internet directories now available. You can also search for newsgroups using the UseNet function of any major search engine.

ADVICE TO THE SCIENTIST-AUTHOR

What has been said above should give you some idea about the various opportunities for publication on the Internet. In fact, let me emphasize again that sending *any* information through the Internet means you are entering it into the realm of public exchange. Keep this in mind; it will help guide your professional conduct in this new and still somewhat unsettled medium.

When looked at with a level eye, Internet publication does not really appear enormously more variable than print publication (if we consider all modes of printed contact among researchers). But it is less certain in status. We still live in a universe dominated by hard copy—this is a practical necessity, given the current state of technology, the structure of our institutions, and how professional life is conducted.

At present, the "safest" mode of Internet publication, particularly for career reasons, is in refereed e-journals. Many of the same practical factors exist here as for print: be sure to consult the information to contributors that specifies length, style, document formatting, submission procedures, and so forth, which you *must* follow for your document to be seriously considered for publication. Most e-journals now also contain specifications for how to reference online material, including e-mail and newsgroup messages. If any of this is ever in doubt, send an e-mail message to the journal (addresses are usually found at the bottom of its home page) asking for clarification.

If you have the means and know-how, and the journal is amenable, submit your material in HTML, as this will make your article much more useful and interesting and flexible, available for future updating and improvement. The journal may, in fact, do this for you. But it is also to your advantage to learn how to do it yourself (a variety of different software editors for HTML can be found at most computer stores, and make the process relatively painless). Some understanding of how to use HTML will place you in the forefront of Web communication, specifically, and will give you certain powers over your own material, including the power to self-publish. Note that authors frequently insert hyperlinks to figures, references, their own e-mail addresses, their institution's home pages, and whatever else a journal might allow. Even more useful, from your reader's point of view, are links to documents in the reference list: being able to jump directly to a full-text version of a cited article is an enormous advance that Internet communication offers. While copyright and economic realities are likely to prevent this from becoming total and universal (a link to every cited reference in every article), it still offers something previously almost inconceivable. Moreover, it is now possible to add a new type of citation list altogether to online articles, consisting of links to papers that have subsequently cited your original. The creation of such online "citational communities" must be considered an exciting new bibliographic capability offered by the Internet.

Though the academic tenure-reward system has not yet fully caught up with reality, e-journals represent a growing piece of the scientific future. As a researcher, don't overlook this new forum as a source for publishing your results.

Even admitting whatever problems and caveats still exist, the online world holds out a bewildering and magnificent array of possibilities for scientific publication. In some ways, it returns us to the origins of the scientific article itself, to a time when a new set of expressive opportunities first appeared on the scene, introducing both order and chaos into the domain of shared knowledge.

CHAPTER FIFTEEN

Dealing with the Press

FACTS AND ISSUES

Some scientists I know love to tell war stories about the media. At conferences and parties, they trade tales of journalistic abuse as if showing scars in a steam room. When pressed a bit, however, these same researchers will also admit to examples of good, even excellent coverage, informative pieces they've seen on other disciplines, magazine articles or documentaries that have treated a familiar topic with both pith and panache. Mixed attitudes, in fact, are common within the scientific community. Indeed they are endemic. Are they, however, inevitable? No, at least not in all cases. A growing number of scientists have begun to appreciate the reasons why the public deserves to be informed about their work. Here are a few.

Scientific knowledge has enormous power in the contemporary world. Yet, unless translated into more ordinary language, this knowledge is all but inaccessible to the vast majority of people, including scientists themselves (if we think of a botanist trying to fathom quantum theory). Most scientific work, meanwhile, is supported by public funds and has certain public goals—to advance knowledge, increase material wealth, support national and institutional prestige, and underwrite professional success, among others. Moreover, science is involved in—nay, often defines—burning realities of immediate social importance: medical research, energy resources, environmental pollution, global warming, biotechnological progress, and many more.

Given these basic truths—power, arcaneness, public

dependence, and social effect—it's only to be expected that science would eventually become the topic of sustained coverage in the press throughout much of the industrialized world. Such, indeed, is the present state of affairs: science is news.

Styles of media attention granted to science have changed considerably since the Second World War. During the 1950s and 1960s, coverage in the press tended to be celebratory: science was the bringer of discoveries and advances; the measure of progress; provider of jobs, national strength, white-coat wizardry. Scientists were often portrayed as heroes or benefactors, working for the good of humankind. The late 1960s and 1970s saw a notable shift in the culture of journalism, however, responding to large-scale political and social trends. Science became a source of potential "risks" and "dangers," as well as "breakthroughs," and a font of elite authority lacking proper accountability. This led to more "investigative" and critical reporting, and to a mixture of favorable and unfavorable scrutiny. As of the 1980s and 1990s, in the wake of the "computer revolution" and other celebrated advances, the pendulum has swung back again, though not entirely. On the one hand, we are again awash in images of *scientia magnificat,* with much of the old language about "miracles" and "breakthroughs" resurrected in full. On the other hand, however, the discourse of risk is still very much with us, and has expanded into new realms (DNA research, irradiated food, etc.).

The biggest change of all, however, from the point of view of the average researcher, has been in the degree of coverage. This has continued to grow. Moreover, it has been fed by calls from various corners for better public understanding of science and for improved "scientific literacy" among the lay populace. Politicians, educators, civic groups, and scientists themselves have all been involved in this advocacy. The result has been greater visibility for science in both print and broadcast media.

Today, science-related stories frequently occur on the front pages of major newspapers (a number of which also have weekly pull-out science sections), in many national magazines, on news broadcasts and documentaries, and elsewhere. Much of this, moreover, is due to the effort of scientists themselves. They have sought to get involved and gain a measure of control over the types of publicity at issue—as experts, as spokespeople, as public debaters, even as personalities. Like it or not, science is very much in the public eye, ear, and mind. And there is every indication that this will continue.

Still, scientists are often divided on how to think about press coverage. Some—an increasing number, in fact—recognize the need for contact with the public and may actively pursue it. Many remain suspicious or uncertain, however, and feel encouraged in this hesitancy by the frenzy they see surrounding such episodes as cold fusion, the failed superconducting supercollider, the Human Genome Project, and the like. Some are openly hostile toward the press and any communication with it. These differences in attitude are only partly generational; there are still no standards taught about how science and the press should interact, what ideals should be met. I suspect a majority of researchers fall somewhere between the first and second categories noted above—they see the advantages, professional and philosophical, of public exposure, but remain a bit wary of what it involves. Highly visible scientists, such as Carl Sagan or Stephen Jay Gould, have attracted criticism from colleagues for their waltzes in the limelight.

The realities of the present situation therefore demand a certain intelligent awareness on the part of any scientist who may one day come in direct contact with media representatives. To simply dismiss science journalism as popular pabulum, or shun it as invasive and debasing, would be very wrong and, indeed, self-defeating. Similarly, viewing the press solely as a platform for promotion—for selling one's own work, department, or field—is also misguided, and a recipe for trouble.

The media define a vast, complex series of institutions. The people who make up these institutions are as mixed in their competence, savvy, and interest as are those of any other, equally sizable "estate"—like science. Even more, journalists work under a range of constraints that *must* be acknowledged and understood, for they determine in part what appears in their reports. To deal with the media effectively and intelligently, therefore, it helps to be aware of certain realities involved in the encounter.

REALITIES OF JOURNALISM

First, let us confront the basic question: is coverage of science necessary in the press? The answer is, absolutely. Honesty demands that we admit the role of science journalism in any democracy that hopes for an informed public.

This role, however, is far from simple. Rendering what is essentially an elite, expert knowledge into terms accessible to Mr. T. C. Mits (The Celebrated Man in the Street) involves much more than merely giving the emperor a new wardrobe. It demands something closer to a *rewriting* of scientific knowledge, adapting it to a wholly different context of presentation and audience. This is no simple achievement, if done well. Moreover, one must add to this challenge the realities of contemporary journalism, which include (but are not limited to) (1) writing on deadline; (2) limitations of space; (3) editorial control; (4) pressures for simplification; (5) demand for definitive statements (from "experts"); and (6) the need to attract interest.

How do reporters cope with this situation? Through a host of choices regarding type of material, level of detail, use of metaphor, and various narrative techniques. Fortunately (for reporters) and sometimes unfortunately (for scientists and the public), many conventions exist for obeying these requirements. When asked, for example, "What makes science news?" one well-seasoned practitioner of the art recently admitted that the list of criteria, in order of weight, would have to include fascination value, size of possible audience, scientific importance, reliability of the results, and timeliness (Rensberger 1998).[13] The significance of this list will not escape most readers here.

Newspeople who write about science and technology range from full-time science journalists to those who cover "the science beat" only occasionally. An increasing number have scientific training, possibly at a graduate school level. Others have migrated into science journalism from political, economic, or feature reporting. The various types of publications and media for which journalists work can involve different audiences. On the one hand, some periodicals are targeted at scientists (professional and amateur) and science watchers, for example, *Science News, Scientific American,* the *New Scientist,* and the news sections of *Science* and *Nature.* These very often—though by no means categorically—present favorable coverage of new work, but have increasingly taken to discussing public consequences and political contexts as well. On the other hand, newspapers like the *New York Times,* the *London Times, Le Monde, Frankfurter Allgemeine Zeitung,* and *Asahi Shinbun* are likely to show mixed styles of coverage, though in recent years the overall tone has been far more pro than con. The broadcast media, television and radio, are most interested in capsule reports that contain some nugget of information on a new discovery, public health issue, controversy, or the like. This branch of the media is, by far, the most

13. For those scientists who might (or should) be interested, it is well worth the time needed to read through this official guide, endorsed by the National Association of Science Writers, to get an idea of how "the other side" lives and works, how they conceive of their craft and its subjects, including scientists.

constrained in terms of space and time and so looks to hook the viewer or listener often more aggressively than do other news outlets.

What about narrative techniques? These have tended to vary over the decades, too, but return to several standard journalistic approaches (science journalism, be assured, is a subset of journalism generally). One of these involves beginning with a "grab," for example, "If Martin Grossbauer is right, we may all be living in the dark by 2050," and then going on to explain and amplify the startling/enigmatic opening. Another technique is to start with a pun or play on words—"Astronomers like to think dark thoughts, particularly when it comes to questions of matter in the universe." The traditional "four W's and the H" method (who, what, where, when, and how) resembles the standard news story. Since the 1990s, meanwhile, it has become fashionable to adopt techniques from feature writing, in which articles tell "a human story" to get the reader involved on a personal level, for example, "Mary Johnsen had never heard of interleukin 2 before she was diagnosed with deadly hepatitis-B."

Such are but a few examples. Obviously, much more could be said about these techniques, and the images of science and scientists they offer. The important thing for researchers, however, is to be aware that the journalist is a craftsperson, who, in the midst of various pressures and constraints, constructs his or her story in particular ways, just as any writer must.

In the end, two great forces pull and tear at the science journalist—the same great forces that rend all of journalism, but that act here with greater intensity. These are the demands to *engage* (win interest, entertain, fascinate) and to *inform* (offer knowledge, increase understanding, urge awareness). These demands take place within an institutional setting where economic realities are often pressing, even paramount. Our capitalistic system of making and selling news means that any such outlet must attract attention—lots of it—to succeed and survive. This is not a negotiable reality.

Ideally, of course, reporters are third-party observers who work consciously in the public interest. In practical, day-to-day terms, however, they must satisfy their editors, whose job it is to sell newspapers, broadcast value, whatever (and thus keep themselves and their reporters employed). The realities of contemporary journalism ensure that science correspondents, however essential to an informed democracy, only occasionally view their main responsibility as educating the public. This is perhaps unfortunate, but largely inevitable, given the way things work. A crucial point for scientists is to avoid the urge to blame reporters for these realities.

A BRIEF COMPARISON

How does the press see its own task with regard to science? How might this compare with the ways in which media work is viewed by those who have studied the subject closely? Let's consider, again, how a well-known, successful science writer describes the trade:

> When I tell a tale, I like to write it as though it were a film . . . creating close-ups, medium shots, and long shots through words, so that readers can visualize the story and feel that they are there, watching. But you can only write those revealing close-ups and backgrounding long shots if you did the reporting that brought you the material to work with. Good writing, with its need for fine detail, eloquent quotes, and vivid imagery, depends on good reporting. There are no shortcuts. Be willing to spend the time it takes to get to know the facts of the story, the characters in the story, and the issues the story raises. (Knudson 1998, 77)

Scientists might take note here that "the facts" are only one thread in the larger tapestry to be woven. There are also "characters" and "issues," "eloquent quotes" and "vivid imagery." It is human interest and social significance, not knowledge alone, that form the pilot subjects of the journalistic effort.

That being said, let us now look at some rather sobering words by a noted sociologist (Dorothy Nelkin), who has herself written eloquently on how the press covers science:

> Journalists convey certain beliefs about the nature of science and technology, investing them with social meaning and shaping public conceptions . . . Was interferon a "magic bullet" or a "research tool"? Was Three Mile Island an "accident" or an "incident"? . . . Is dioxin a "doomsday chemical" or a "potential risk"? . . . Are incidents of scientific fraud "inevitable" or "aberrant"? Some words imply disorder or chaos; others certainty and scientific precision. Selective use of adjectives can trivialize an event or render it important; marginalize some groups, empower others; define an issue as a problem or reduce it to a routine. (Nelkin 1995, 11)[14]

14. This work should be required reading for every scientist. Also very good on science and the media are the essays collected in Scanlon, Whitelegg, and Yates 1998.

In other words, the press does not merely tell tales (tall, short, or otherwise). It also acts as political broker and agenda maker. Thus, for scientists it helps greatly to read press accounts of scientific work with a critical eye, noting how material is ordered, what images are used to present it, what levels of sophistication are assumed (or not assumed). Science writers have their own considerable power to shape public sensibility and establish menus for public policy. But they are not alone in this process, by any means. Consider their sources.

THE MOTIVES OF SCIENTISTS

As a species, scientists are no longer categorically shy or passive in the face of media attention. An increasing number have transformed themselves into spokespeople, public debaters, even regular commentators. There is also the demand, brought to bear by many sources—scientific societies and research institutions among them—that investigators should actively reach out to the public, make their work comprehensible, available. Whether this means education, promotion, or popularization, however, often depends on local circumstances. The scientist of today does not so much descend from the gates of Olympus, as emerge into the spotlight from less elevated precincts.

It does not require a public policy expert to suggest that positive exposure can lead to many benefits—increased influence, enhanced funding, job opportunities, an advance into favor for one's own discipline. Conversely, negative publicity can be very damaging to a program, career, or area of research. Scientists today realize that the stakes involved in media coverage can be very high. Science journals, too, are very aware of the advantages media publicity can bring. Some (e.g. *Science,* the *New England Journal of Medicine*) provide advance copies of their upcoming issues to journalists, who then cover stories and articles in them, making such journals appear the key sources of new science.

This has certainly worked in a few cases, to mutual benefit for researchers and reporters. But it can easily backfire, too. As already noted, the press has its own needs and agendas, and experienced journalists have their antennae up for attempts to exploit them. As the list of news criteria given earlier shows, it is a mistake to think that positive coverage, replete with words like "revolutionary" and "discovery," is the mark of uncritical journalism: this type of sensationalism (let us be frank: how many revolutions and breakthroughs can there be?) is very much what the press has decided to sell to the public, though often with certain cautions attached. A problem, however, is that reality seldom accedes to hype. When the relevant "miracles" fail to recur or to produce the predicted results, reporters will usually turn negative or mute.

The complexity of the situation, and the stakes involved, have led many research institutions to put on staff a science press officer or media consultant. This person is a notable (if still local) addition to the world of media interaction. He or she can act as an adviser, middleman, filter, or mediator between scientists and journalists—that is, in whatever ways seem needed to make contact with the press smooth, efficient, and beneficial to both parties. The results are not perfect; moreover, scientists may still seek to initiate contact on their own, and journalists will always need to speak to their sources directly.

Let us admit that, as scientists, we too are torn by competing forces. These, to be sure, are not the same as those for the media. We want favorable coverage for the labor of our hearts and minds, a clean and shining public image. Yet our training, and the ethics of our profession, tend to make us suspicious and even defensive toward rapt public exposure. Much of scientific work is aimed at establishing competitive advantage, being at the forefront of a particular area, and scrutinizing others' work with a critical eye. The mentality that goes with this is one of caution, keeping things quiet and proprietary until published, distrusting large claims offered without supporting detail. Moreover, scientists are trained to view themselves as experts, as being separate from the nonscientific public in particular ways. We usually see our own knowledge as exciting, valuable, even admirable—otherwise, what's the point?—whereas to the press, it can qualify as mere "subject matter" or "copy" *unless* filled out, even dramatized, by other elements. The realization of this truth can be disillusioning.

Scientific publication and media publication are therefore two very different creatures. One frequent problem for researchers is that they tend to measure the second by the standards of the first (reporters are not scientists!). Similarly, it is rarely a good idea to "market" one's work in total capitulation to media demands (scientists are not press agents!). The greater danger, in fact, may lie exactly here: once given the spotlight and eager to speak on behalf of their field, scientists may fall into the trap of playing the press's own game, that is, speaking a language that is distinctly not their own. I've mentioned the downside of taking up the tired and misleading discourse of "breakthrough," "revolution," "miracles," and so on. No less is this the case for the speech of "perils," "threats to humanity," "disaster for science," and the like. Prophetic optimism and dark despair are not the stuff of intelligent comment. It is the scientists' eminent task to negotiate these realities without falling prey to them entirely.

THE PUBLIC'S INTEREST

The public is interested in scientific work for a great variety of reasons. Science is powerful, expensive, elitist, inaccessible, yet also forward looking, optimistic, full of promise, even, at times, spectacular. More than any other area of knowledge, science carries with it the sense of advancement, moving ahead, exploration, newness. Science visibly improves its own powers, adds to itself, and carries us all forward, in general feeling, with it. There have long been moral and emotional reasons to be informed about the "latest advances," and now there are political and social ones as well. Knowing some of the facts and issues surrounding genetic testing or human cloning allows one to be part, and feel part, of the decision-making process. Some of this knowledge—and related impressions—have come from media reporting itself, past and present. But this does not mean that scientists are required to adopt and repeat such images in every instance.

What does the public need to know about a particular branch of scientific work? There is no simple answer. Indeed, the question itself is often misinterpreted. Researchers, that is, can all too readily confuse public *understanding* of science with public *appreciation*. Understanding (e.g. how nuclear energy is generated) can lead to queries, to criticism, and even to rejection. To know something of science is not necessarily to love it: the truly aware researcher must realize this and be prepared for it.

Interest in science is also deeply affected by medium. Consider that the great majority of media publications are meant to be skimmed, not studied: readers are able to retain very little specific information from a newspaper article, magazine story, or (especially) television or radio broadcast. This lack of retention is due both to the style of exposure (quick, one-time reading or listening) and to the fact that there are usually many such exposures on a wide range of subjects (politics, international events, economics, features, etc.) to be ingested at a single sitting. The topical press is not something that provides people with opportunities for concentrated learning or continuing education. Reporters know this; they know they must write stories, not primers.

Public interest in science is complex, multifaceted, difficult to define in any precise way. But one thing can be said for sure: in the popular press, this interest always comes back, sooner or later, to "news"—with everything that this encompasses. In large part, it includes only the tip of the scientific pyramid—that part of science which is today in progress, being conducted, in the here and now. This is the most provisional—debated, competitive, uncertain—and, in a social sense, exciting part of science. But it is also the most difficult to write about in any definitive way. It is one thing to review for a public audience the basic principles of chemistry, which can be taken from the established literature; it is quite another to discuss the merits, hotly debated, of a new hypothesis on the physical chemistry of superconducting materials.

For scientists, it helps to distinguish between what is involved in knowing, and knowing *about,* a particular subject. Reporters write, and readers expect, stories *about* scientific ideas, advances, effects, and so forth. Technical knowledge per se is not the goal; the aim instead is to communicate something of the character, importance, and implications of this knowledge, what it might signify to the individual reader or viewer. This does not in the least excuse inaccurate reporting, sensationalism, or other common faults of the media. But it does help put boundaries around what scientists might anticipate from press coverage and public interest.

BEING RESPONSIBLE: SOME RECOMMENDATIONS

What is the scientist's primary role vis-à-vis the press? Most often, it will involve acting as an "expert source." What does this mean? In general terms, it means speaking as an inside witness to your own work, and very often to the state of knowledge, effort, and debate in your specific area. Your main duty, in other words, is to communicate the central points of your research, the history and limits of your knowledge, and what it might mean for your own field. Occasionally, you may be called upon to speculate about things beyond your area of expertise—larger social impacts, the future of the discipline, even political trends or the state of science generally.

Put this way, it all sounds pretty simple. But it can be confusing if you are unprepared and find yourself unsure about how to act, what sort of tone to adopt, or how much to "reveal." If you work for an institution with a public relations office, especially one with a science officer or media consultant on staff, it is a good idea to discuss strategy with this person before giving an actual interview, generating a press release, serving on a public panel, or the like.

Reporters are usually very interested in cooperative contact. As a source, you are an asset to them, and may be so again in the future. A journalist would much prefer to discover and cultivate a spring than poison the well. Nonetheless, for the sake of producing an acceptable (and informative) article, they may need to ask questions that appear to be critical or challenging. A list of such queries might include some or all of the following:[15]

- How do you know what you claim to know? Is this speculation or is it based on actual research? (This question might be asked in cases where broad claims with social importance are made.)
- How and where did you do your research? What basic approach did you take? What methods did you use? How orthodox or unorthodox were these methods?
- Are your results reproducible? How consistent have they been between different studies?
- How accurate are your data? What level of uncertainty is involved?
- How sure can you be about your conclusions or interpretations? What alternative explanations might be given for your data?
- What do other researchers think about this topic? How much debate or controversy is there within the field? Who disagrees with you, and why?
- When did you first become interested in this topic, and are there any personal reasons related to your background, or that of your friends or family, that led you to it? (This question will usually be approached subtly.)
- If your ideas prove correct, how might you profit from them, personally and professionally? (Again, indirect inquiries may try to get at this information.)
- Are any potentially negative issues or effects associated with your work?

All these questions may come up, for example, in an interview or a series of ongoing discussions with one or more reporters. The well-prepared scientist will be able to answer them with a significant degree of honesty and without feeling he or she is under attack or is being asked to dance through a minefield.

The following recommendations are offered to help you, the scientist, deal with such questions, as well as others that might arise.

Whether you're due for an interview or not, keep a critical eye tuned to media coverage of science, especially in your field. Look at how a

15. A similar, though more abbreviated, list is provided by Victor Cohn in his perceptive article "Coping with Statistics" (Blum and Knudson 1998, 102–109).

particular subject is handled, what sorts of questions are being addressed, the type of language the reporter uses, how detailed is the handling of the relevant knowledge. You might even look for patterns in the styles of coverage given your discipline—what sorts of metaphors or images are regularly employed.

When you see an excellent article, report, or documentary, save it for study. Analyze what it is you like about it (is it the language, the depth of coverage, the organization, the pacing . . . all of the above?). Adopt what you can from it for discussing your own work.

During any interview situation, find out how much science the reporter knows, what scientific training she or he has had, and how much background reading she or he has done on the particular topic at issue. This will give you an idea of how complex or basic your answers to questions need to be. Two of the most common errors scientists make in interviews are to condescend to reporters and to use too much jargon.

Beware of being cast as an Expert. Journalists sometimes have a tendency to turn scientists into authorities on matters well outside their area of professional competence, for example, as spokespeople on science generally, or on religion, political events, social movements, and other fields (e.g. physicists talking about the nature of the human mind, evolution, literature, philosophy, etc.). Try to avoid this as much as possible; otherwise, you risk making misleading, even ignorant statements and sounding foolish in front of those who have devoted their own professional careers to study of the relevant subject.

Also, take care not to play the "media game" too avidly, that is, indulging in exaggerated speech, clichés that will appeal "to the masses," and so forth. This may make you look colorful, but it stands a good chance of providing poor-quality information and cheapening your stance in front of your colleagues.

Similarly, to the degree possible, don't pander to the general tendency of the press to speak in terms of "miracles," "revolutions," "perils," "magic bullets," and so on. Such language most often provides a public disservice in terms of characterizing the actual significance of a finding, hypothesis, or conclusion.

Accept the reality that, in dealing with the press, scientific material must be simplified, often recast for the nonscientific audience. Think carefully, if you have time, about how to do this.

Remember that your dealings with the press involve human relationships; they are not merely institutional. To express open suspicion and distrust toward reporters is to effectively burn connections that you might otherwise profit from. Remember, too, that reporters and scientists share a central effort: to inform others. They are both in the business of writing, of shaping their material into a coherent set of messages.

Be prepared to be asked about the social implications of your work. Reporters often want scientists to predict what the future might look like. Be clever; offer several alternatives. This will keep you from being held to any particular scenario later on.

If there is controversy or debate about the topic at hand, acknowledge this up front. Cast it in a positive light, in the sense that critical review and scrutiny of new ideas are how science progresses.

Be ready to point out the limits to your knowledge or that of your field. In this regard, try to avoid the "we don't know" type of response. In most cases, you are being consulted as an expert; therefore speak in positive terms if you can, for example, about "questions now being pursued," "new areas of work opening up."

Do not—I repeat, do *not*—expect any article, broadcast, report, or commentary to be 100% accurate in every detail. We do not even expect this in scientific articles or in laboratory work generally, so it is unfair to set it as a standard for the press. Rank any errors you do see in terms of their real importance to the story being told (i.e. in an article about stellar evolution, a mistake in the description of big bang theory will be more significant than a blunder in stating the total number of galaxies, even if counting these has been involved in your own work for the last two years).

Last, but by no means least, remember that, as a source, you cannot dictate in any final way what will be printed or said. You may be allowed to review for accuracy relevant copy or recorded video, but this does not give you editorial purview or censorship rights over anything else (opinions, language, etc.). It may help in this regard to reflect on the following:

> For the scientist who wants to be a source of science news, forewarned is forearmed: collaborating with journalists and adapting to journalistic conventions may give scientists more, rather than less, control over the emphasis and tone of the resulting story. But the last word will always go to the journalists, because science journalism is much more about journalism than it is about science. (Gregory and Miller 1998, 130)

CHAPTER SIXTEEN

In Conclusion

I began this book with the truth that research and communication form a continuum. In the end, it is more accurate to say that, in the real world of daily scientific work, they are inseparable. Research involves a number of central activities, and communicating is one of them. Using words is interwoven with laboratory work, library work, theoretical work, collegial contact, and every other type of labor associated with science, from beginning to end. If you can't communicate adequately with your peers, you can't do research, at least in any truly productive sense. Poor writing makes for bad science.

Such, at least, is how a stern parent might put the matter. I prefer to say, more gently (but no less realistically), that the sharing of knowledge—especially in formal, communal ways—is the nutritive process that makes the body of science a living, growing enterprise. Reading, speaking, and writing, with at least a functional level of skill, thus become the inevitable responsibility of every scientist who wishes to contribute directly to the vitality of that corpus. To present one's work to others therefore need not be an act of mere survival. It can be a conscious deepening of one's participation in a domain that, after all, one has chosen for a lifetime of effort and loyalty.

Contrast this with the standard idea that the scientist must "write up" her or his research. What is meant by this little phrase, so often used, so rarely questioned? Partly, it recalls the dreaded lab report of school science, with its demand to get required data down on paper before "writing up" the results (in apparent seesaw fashion). Something of this carries over into professional work, where this phrase suggests the investigator must first call a halt to *real* work—in the lab, the field, the office, wherever—before beginning to write. He or she must then sit down, in monastic manner, draw breath, and wrestle the beast of language into submission. Writing thus appears an unfortunate, even lamentable obstacle, an intrusive obedience to outside demands.

Such attitudes, of course, are fateful. Shall we say destructive, self-fulfilling? Writing is work, certainly, just as any process of experimentation and discovery must be—but no harder than this, and no less significant.

Writing does involve a journey to the interior, and for many, this may not be an especially pleasant experience. There can be significant anxiety about giving expression to one's work, for, in truth, it is a type of exposure. Yet I wager that if we go back far enough in our own private histories, to when we first felt attracted to science as lifework, we will find embedded in our nascent and possibly cinematic images of "science" itself a certain desire to influence others, to compete and ascend, and to connect with a larger world to which we would add something important. The hope of such addition, and the ambitions it nucleated, rested on the sense of making our mind and work available to others.

Like any author, scientists add themselves to the world through symbols; this is how they make their contribution. The scientist today has a greater, more exciting range of expressive outlets to employ than ever before. It is, indeed, an exciting and demanding era. There are more forms of publication to be aware of, more media to learn and take advantage of, more avenues of contact among professionals and institutions. Science is much larger than it once was, more urgent, diverse, complex, hungry. But communication remains its core substance. More than ever, scientists are scholars of the written word. Knowing how to write with a modicum of functional skill or better, and even a degree of pride or pleasure, gives one of the most essential endowments for negotiating successfully the demands of the scientific life.

It has been my effort in this book to outline ways that will help the student and practitioner of science acquire or advance this endowment. A principal theme has been the use of models of good writing as a basis for improving one's own expression. This is hardly an original idea; indeed, it is among the most ancient of wisdoms. But it requires time, and therefore patience. We continue to need good writers to keep our science vital and growing, and if we cannot train them fully during their school years, we need to provide methods for them to develop thereafter. Knowingly or not, all writers learn by example, and by experiment. This is a sine qua non of any art or craft. Making this process conscious, even methodical, can therefore grant one considerable advantage, whether as an apprentice or self-styled journeyman.

"The world is a noisy business," said Daniel Defoe, one of the most prolific authors who ever put pen to paper (some 560 books, pamphlets, and journals came from his hand). If we are to count writing in general a major contribution to this noise, then certainly science is the source of untold reverberations that carry us all from the quieting past into an opening future. Modern science largely began as literature—the sharing of knowledge and experience through the publication of books, journals, diaries, translations, and more. Literary it may no longer be, but literature, in the larger sense, it certainly remains.

SELECTED BIBLIOGRAPHY

Adams, D. D., and M. Naguib. 1999. "Carbon gas cycling in the sediments of Plußsee, a northern German eutrophic lake, and 16 nearby water bodies of Schleswig-Holstein." *Archiv für Hydrobiologie,* Special Issue on Advances in Limnology, 54:91–104.

Alley, M. 1996. *The Craft of Scientific Writing.* 3rd edition. New York: Springer-Verlag.

Anholt, R. H. 1994. *Dazzle 'em with Style: The Art of Oral Scientific Presentation.* New York: W. H. Freeman.

Bainbridge, D. 2001. Review of *The Karoo: Ecological Patterns and Processes* by W. R. J. Dean and S. J. Milton. *Ecoscience* 8, no. 1:139.

Barolo, S., R. G. Walker, A. D. Polyanovsky, G. Freschi, T. Keil, and J. W. Posakony. 2000. "A notch-independent activity of suppressor of Hairless is required for normal mechanoreceptor physiology." *Cell* 103:957–969.

Bess, T. D., A. B. Carlson, C. Mackey, F. M. Denn, A. Wilber, and N. Ritchey. 2000. "World Wide Web access to radiation datasets for environmental and climate change studies." *Bulletin of the American Meteorological Society* 81, no. 11:2645–2652.

Bishop, C. T. 1984. *How to Edit a Scientific Journal.* Baltimore: Williams and Wilkins.

Blum, D., and M. Knudson, eds. 1998. *A Field Guide for Science Writers.* New York: Oxford University Press.

Booth, Vernon. 1993. *Communicating in Science.* 2nd edition. Cambridge: Cambridge University Press.

Borowick, J. N. 1996. *Technical Communication and Its Applications.* New York: Prentice Hall.

Briscoe, M. Y. 1996. *Preparing Scientific Illustrations: A Guide to Better Posters, Presentations, and Publications.* 2nd edition. New York: Springer-Verlag.

Carlson, R. W., et al. 1999. "Hydrogen peroxide on the surface of Europa." *Science* 283, no. 5410:2062–2063.

Chambers, J. S. 1983. *Graphic Methods for Data Analysis.* London: Chapman and Hall.

Cleveland, W. S. 1993. *Visualizing Data.* Summit, NJ: Hobart Press.

———. 1994. *The Elements of Graphing Data.* Revised edition. Boca Raton, FL: CRC Press.

Cochran, W. 1984. *Geowriting: A Guide to Writing, Editing, and Printing in Earth Science.* 4th edition. Alexandria, VA: American Geological Institute.

Council of Biology Editors, Style Manual Committee. 1994. *Scientific Style and Format: The CBE Manual for Authors, Editors, and Publishers.* 6th edition. Cambridge: Cambridge University Press.

Crystal, D. 1995. *The Cambridge Encyclopedia of the English Language.* Cambridge: Cambridge University Press.

Cui, L., A. I. Soldevila, and B. A. Webb. 2000. "Relationships between polydnavirus gene expression and host range of the parasitoid wasp *Campoletis sonorensis.*" *Journal of Insect Physiology* 46:1397–1407.

Cui, Y., and C. M. Lieber. 2001. "Functional nanoscale electronic devices assembled using silicon nanowire building blocks." *Science* 291, no. 5505:851–853.

Curie, M. 1952. "Radio-active substances." In H. M. Leicester and H. S. Klickstein, eds., *A Source Book in Chemistry, 1400–1900,* 522–531. Cambridge, MA: Harvard University Press.

Dahl, R. 1953. *Someone Like You.* New York: Knopf.

Davis, M. 1997. *Scientific Papers and Presentations.* San Diego, CA: Academic Press.

Day, R. A. 1995. *Scientific English: A Guide for Scientists and Other Professionals.* 2nd edition. Phoenix, AZ: Oryx Press.

———. 1998. *How to Write and Publish a Scientific Paper.* 5th edition. Phoenix, AZ: Oryx Press.

Dear, P., ed. 1991. *The Literary Structure of Scientific Argument.* Philadelphia: University of Pennsylvania Press.

Dearborn, D., G. Raffelt, P. Salati, J. Silk, and A. Bouquet. 1990. "Dark matter and the age of globular clusters." *Nature* 343, no. 6256:347–348.

Dickstein, A. J., S. Erramilli, R. E. Goldstein, D. P. Jackson, and S. A. Langer. 1999. "Labyrinthine pattern formation in magnetic fluids." *Science* 261, no. 5124: 1012–1015.

Dodd, J. S., ed. 1998. *The ACS Style Guide: A Manual for Authors and Editors.* 2nd edition. New York: American Chemical Society.

Fourdrinier, S., and H. J. Tichy. 1988. *Effective Writing for Engineers, Managers, Scientists.* 2nd edition. New York: John Wiley and Sons.

Fuller, S. 1997. *Science.* Manchester, England: Open University Press.

Geikie, A. 1970. "On denudation now in progress." In K. F. Mather and S. L. Mason, eds., *A Source Book in Geology, 1400–1900,* 523–528. Cambridge, MA: Harvard University Press.

Gill, D. 1985. "Depositional facies of Middle Silurian (Niagaran) pinnacle reefs, Belle River Mills Gas Field, Michigan Basin, southeastern Michigan." In P. O. Roehl and P. W. Choquette, eds., *Carbonate Petroleum Reservoirs,* 123–139. New York: Springer-Verlag.

Ginsparg, P. 1996. "Winners and losers in the global research village." Joint ICSU Press/UNESCO Conference on Electronic Publishing in Science, UNESCO, Paris, 19–23 February. Available at http://associnst.ox.ac.uk/confproc.htm.

Goodlad, S. 1996. *Speaking Technically: A Handbook for Scientists, Engineers, and Physicians on How to Improve Technical Presentations.* London: World Scientific Publishing Co.

Gopen, G. D., and J. A. Swan. 1990. "The science of scientific writing." *American Scientist* 78:550–558.

Gregory, J., and S. Miller. 1998. *Science in Public: Communication, Culture, and Credibility.* New York: Plenum.

Guber, A., X. Su, M. Kanamitsu, and J. Schemm. 2000. "The comparison of two merged rain gauge–satellite precipitation datasets." *Bulletin of the American Meteorological Society* 81, no. 11:2631–2644.

Hailman, J. P., and K. B. Strier. 1997. *Planning, Proposing, and Presenting Science Effectively.* Cambridge: Cambridge University Press.

Hazuda, D. J., P. Felock, M. Witmer, A. Wolfe, K. Stillmock, J. A. Grobler, A. Espeseth, L. Gabryelski, W. Schleif, C. Blau, and M. D. Miller. 2000. "Inhibitors of strand transfer that prevent integration and inhibit HIV-1 replication in cells." *Science* 287, no. 5453:646–649.

Heaton, T. H., and S. H. Hartzell. 1986. "Source characteristics of hypothetical subduction earthquakes in the northwestern United States." *Bulletin of the Seismological Society of America* 76, no. 3:675–708.

Hers, H.-G. 1984. "Making science a good read." *Nature* 307, no. 5256:205.

Hiraguchi, T., and T. Yamaguchi. 2000. "Escape behavior in response to mechanical stimulation of hindwing in cricket, *Gryllus bimaculatus.*" *Journal of Insect Physiology* 46:1331–1340.

Hodges, E. R. S. 1988. *The Guild Handbook of Scientific Illustration.* New York: John Wiley and Sons.

Holmes, A. 1965. *Principles of Physical Geology.* New York: Wiley.

Hoover, H. 1980. *Essentials for the Scientific and Technical Writer.* 2nd edition. Mineola, NY: Dover.

Huiskes, R., R. Ruimerman, G. van Lenthe, and J. D. Janssen. 2000. "Effects of mechanical forces on maintenance and adaptation of form in trabecular bone." *Nature* 405, 8 June, 704–706.

Huth, E. J. 1990. *How to Write and Publish Papers in the Medical Sciences.* 2nd edition. Philadelphia: Lippincott Williams and Wilkins.

Huxley, L., ed. 1903. *Life and Letters of Thomas Huxley.* Vol. 1. London: Macmillan and Co.

International Council for Science (ICSU) Press Committee on Dissemination of Scientific Information. 2001. *Guidelines for Scientific Publishing.* Available at http://associnst.ox.ac.uk/icsuinfo/guidelines.htm.

Iverson, C., ed. 1997. *American Medical Association Manual of Style: A Guide for Authors and Editors.* Baltimore: Williams and Wilkins.

Jordan, C. F., Jr., and J. L. Wilson. 1994. "Carbonate reservoir rocks." In L. B. Magoon and W. G. Dow, eds., *The Petroleum System: From Source to Trap,* 141–158. American Association of Petroleum Geologists Memoir 60. Tulsa, OK: American Association of Petroleum Geologists.

Kenny, P. 1982. *Public Speaking for Scientists and Engineers.* Bristol, United Kingdom: Adam Hilger.

Knudson, M. 1998. "Telling a good tale." In D. Blum and M. Knudson, eds., *A Field Guide for Science Writers,* 77–85. Oxford: Oxford University Press.

Lannon, J. 1979. *Technical Writing.* Boston: Little, Brown and Co.

Levine, G., ed. 1987. *One Science: Essays in Science and Literature.* Madison: University of Wisconsin Press.

Lindahl, T., and R. D. Wood. 1999. "Quality control by DNA repair." *Science* 286, no. 5446:1897–1905.

Lindsay, D. 1995. *A Guide to Scientific Writing.* 2nd edition. New York: Longman.

Locke, D. 1992. *Science as Writing.* New Haven, CT: Yale University Press.

Mann, S., D. D. Archibald, J. M. Didymus, T. Douglas, B. R. Heywood, F. C. Meldrum, and N. J. Reeves. 1993. "Crystallization at inorganic-organic interfaces: Biominerals and biomimetic synthesis." *Science* 251, no. 5126:1286–1292.

Matthews, J. R., J. M. Bowen, and R. W. Matthews. 1996. *Successful Scientific Writing: A Step-by-Step Guide for Biomedical Scientists.* London: Cambridge University Press.

Medawar, P. B. 1990. "Is the scientific paper a fraud?" In *The Threat and the Glory,* 228–233. New York: HarperCollins.

Michaelson, H. B. 1990. *How to Write and Publish Engineering Papers and Reports.* 3rd edition. Phoenix, AZ: Oryx Press.

Michelson, A. A., and E. W. Morley. 1887. "On the relative motion of the earth and luminiferous ether." *American Journal of Science,* series 3, 34, no. 203: 333–345. Reprinted in W. F. Magie, ed., *A Source Book in Physics,* 369–377. Cambridge, MA: Harvard University Press, 1965.

Mitchell, J. 1970. "The Earth composed of regular and uniform strata." In K. F. Mather and S. L. Mason, eds., *A Source Book in Geology, 1400–1900,* 80–87. Cambridge, MA: Harvard University Press.

Montgomery, S. L. 1996. *The Scientific Voice.* New York: Guilford.

Moreo, A., S. Yunoki, and E. Dagotto. 1999. "Phase separation scenario for manganese oxides and related materials." *Science* 283, no. 5410:2034–2039.

Moretto, L. A., and R. S. Blicq. 1995. *Writing Reports to Get Results: Quick, Effective Results Using the Pyramid Method.* 2nd edition. New York: Wiley and IEEE Press.

Moriarty, M. F. 1997. *Writing Science through Critical Thinking.* Boston: Jones and Bartlett.

Munster, U., E. Heikkinen, M. Likolammi, M. Jarvinen, K. Salonen, and H. De Haan. 1999. "Utilisation of polymeric and monomeric aromatic and amino acid carbon in a humic boreal forest lake." *Archiv für Hydrobiologie,* Special Issue on Advances in Limnology, 54:105–134.

Murray, D. K., and S. D. Schwochow. 1997. "Coalbed gas development in the Rockies: Analogues for the world." In E. B. Coalson, J. C. Osmond, and E. T. Williams, eds., *Innovative Applications of Petroleum Technology,* 31–46. Boulder, CO: Rocky Mountain Association of Geologists.

Nelkin, D. 1995. *Selling Science: How the Press Covers Science and Technology.* New York: W. H. Freeman.

Newton, I. 1958. "New theory about light and colours." In I. B. Cohen, ed., *Isaac Newton's Papers and Letters on Natural Philosophy,* 47–59. Cambridge, MA: Harvard University Press.

Nolting, F., A. Scholl, J. Stöhr, J. W. Seo, J. Fompeyrine, H. Slegwart, J.-P. Locquet, S. Anders, J. Lüning, E. E. Fullerton, M. F. Toney, M. R. Scheinfein, and H. A. Padmore. 2000. "Direct observation of the alignment of ferromagnetic spins by antiferromagnetic spins." *Nature* 405, 15 June, 767–769.

Nosé, M., S. Ohtani, A. T. Y. Lui, S. P. Christon, R. W. McEntire, D. J. Williams, T. Mukai, Y. Saito, and K. Yumoto. 2000. "Change of energetic ion composition in the plasma sheet during substorms." *Journal of Geophysical Research* 105, no. A10:23277–23286.

O'Connor, M. 1991. *Writing Successfully in Science.* New York: HarperCollins.

Paradis, J. G. 1987. "Montaigne, Boyle, and the Essay of Experience." In George Levine, ed., *One Science: Essays in Science and Literature,* 59–91. Madison: University of Wisconsin Press.

Paradis, J. G., and M. L. Zimmerman. 1997. *The MIT Guide to Science and Engineering Communication.* Cambridge, MA: MIT Press.

Pechenik, J. A. 1993. *A Short Guide to Writing about Biology.* 2nd edition. New York: HarperCollins.

Porush, D. 1996. *A Short Guide to Writing about Science.* New York: Addison-Wesley.

Price, D. J. de Solla. 1963. *Little Science, Big Science.* New York: Columbia University Press.

Priestley, J. 1952. "Of dephlogisticated air, and of the constitution of the atmosphere." In H. M. Leicester and H. S. Klickstein, eds., *A Source Book in Chemistry, 1400–1900,* 113–123. Cambridge, MA: Harvard University Press.

Rensberger, B. 1998. "Covering science for newspapers." In D. Blum and M. Knudson, eds., *A Field Guide for Science Writers,* 7–16. New York: Oxford University Press.

Rhodes, R. 1995. *How to Write: Advice and Reflections.* New York: Quill.

Santen, L., and W. Krauth. 2000. "Absence of thermodynamic phase transition in a model glass former." *Nature* 405, 1 June, 550–551.

Scanlon, E., R. Hill, and K. Junker. 1999. *Communicating Science: Professional Contexts.* London: Routledge, in association with the Open University.

Scanlon, E., E. Whitelegg, and S. Yates. 1998. *Communicating Science: Contexts and Channels.* London: Routledge, in association with the Open University.

Schoenfeld, R. 1989. *The Chemist's English.* New York: John Wiley and Sons.

Shortland, M., and J. Gregory. 1991. *Communicating Science: A Handbook.* New York: John Wiley and Sons.

Sides, C. S. 1991. *How to Write and Present Technical Information.* Phoenix, AZ: Oryx Press.

Slade, C. 1994. *Form and Style: Research Papers, Reports, Theses.* 9th edition. New York: Houghton Mifflin.

Sprat, T. 1667. *The History of the Royal Society of London, for the Improving of Natural Knowledge.* Reprint. Ed. J. L. Cope and H. W. Jones. St. Louis, MO: Washington University Press, 1958.

Stapleton, P. 1987. *Writing Research Papers: An Easy Guide for Nonnative English Speakers.* Canberra: Australian Centre for International Agricultural Research.

Tsodyks, M., et al. 1999. "Linking spontaneous activity of single cortical neurons and the underlying functional architecture." *Science* 286, no. 5446:1722–1728.

Tufte, E. R. 1983. *The Visual Display of Quantitative Information.* Cheshire, CT: Graphics Press.

———. 1990. *Envisioning Information.* Cheshire, CT: Graphics Press.

———. 1997. *Visual Explanations.* Cheshire, CT: Graphics Press.

Valiela, I. 2000. *Doing Science: Design, Analysis, and Communication of Scientific Research.* Oxford: Oxford University Press.

Warkentin, K. M. 2000. "Environmental and developmental effects on external gill loss in the red-eyed tree frog, *Agalycnis callidryas.*" *Physiological and Biochemical Zoology* 73, no. 5:557–565.

Watson, J. D., and F. H. C. Crick. 1953. "A structure for deoxyribose nucleic acid." *Nature* 171, no. 4356:737–738.

Wilkinson, A. M. 1991. *The Scientist's Handbook for Writing Papers and Dissertations.* New York: Prentice Hall.

Williams, J. M. 1995. *Style: Toward Clarity and Grace.* Chicago: University of Chicago Press.

Wolff, R. S., and L. Yeager. 1993. *Visualization of Natural Phenomena.* New York: Springer-Verlag.

Wood, P. 1994. *Scientific Illustration: A Guide to Biological, Zoological, and Medical Rendering Techniques, Design, Printing, and Display.* 2nd edition. New York: John Wiley and Sons.

Woodford, F. P. 1967. "Sounder thinking through clearer writing." *Science* 156, no. 3776:743–745.

Worsley, D., and B. Mayer. 1989. *The Art of Science Writing.* Philadelphia: Teachers and Writers Collaborative.

Yang, Jen Tsi. 1995. *An Outline of Scientific Writing for Researchers with English as a Foreign Language.* Singapore: World Scientific Publishing Co.

Zimmerman, D. E., and D. G. Clark. 1987. *The Random House Guide to Technical and Scientific Communication.* New York: Random House.

Zinsser, W. 1985. *On Writing Well: The Classic Guide to Writing Nonfiction.* 6th edition. New York: Harper Reference.

Zoback, M. L., M. D. Zoback, J. Adams, M. Assumpcao, S. Bell, E. A. Bergman, P. Bluemling, D. Denham, J. Ding, K. Fuchs, S. Gregersen, H. K. Gupta, K. Jacob, P. Knoll, M. Magee, J. L. Mercier, B. C. Muller, C. Paquin, O. Stephansson, A. Udias, and Z. H. Xu. 1990. "Global patterns of intraplate stress: A status report on the world stress map project of the International Lithosphere Program." *Nature* 341, no. 6240:291–298.

INDEX